Lecture Notes in Mathematics 1533

Editors:
A. Dold, Heidelberg
B. Eckmann, Zürich
F. Takens, Groningen

T0255408

Peter Jipsen   Henry Rose

Varieties of Lattices

Springer-Verlag

Berlin Heidelberg New York
London Paris Tokyo
Hong Kong Barcelona
Budapest

Authors

Peter Jipsen
Department of Mathematics
Iowa State University
Ames, Iowa 50011, USA

Henry Rose
Department of Mathematics
University of Cape Town
7700 Rondebosch
Cape Town, South Africa

Mathematics Subject Classification (1991): 06B05, 06B10, 06B20, 06B25, 06C05, 06C20, 08B05, 08B10, 08B15, 08B20, 08B25, 08B26, 08B30, 03C05, 03C07, 03C20

ISBN 3-540-56314-8 Springer-Verlag Berlin Heidelberg New York
ISBN 0-387-56314-8 Springer-Verlag New York Berlin Heidelberg

© Springer-Verlag Berlin Heidelberg 1992
Printed in Germany

Typesetting: Camera ready by author
Printing and binding: Druckhaus Beltz, Hemsbach/Bergstr.
46/3140-543210 - Printed on acid-free paper

Synopsis

An interesting problem in universal algebra is the connection between the internal structure of an algebra and the identities which it satisfies. The study of varieties of algebras provides some insight into this problem. Here we are concerned mainly with lattice varieties, about which a wealth of information has been obtained in the last twenty years.

We begin with some preliminary results from universal algebra and lattice theory. The second chapter presents some properties of the lattice of all lattice subvarieties. Here we also discuss the important notion of a splitting pair of varieties and give several characterizations of the associated splitting lattice. The more detailed study of lattice varieties splits naturally into the study of modular lattice varieties and nonmodular lattice varieties, dealt with in the third and fourth chapter respectively. Among the results discussed there are Freese's theorem that the variety of all modular lattices is not generated by its finite members, and several results concerning the question which varieties cover a given variety. The fifth chapter contains a proof of Baker's finite basis theorem and some results about the join of finitely based lattice varieties. Included in the final chapter is a characterization of the amalgamation classes of certain congruence distributive varieties and the result that there are only three lattice varieties which have the amalgamation property.

Acknowledgements

The second author acknowledges the grants from the University of Cape Town Research Committee and the South African Council for Scientific and Industrial Research.

We dedicate this monograph to our supervisor Bjarni Jónsson.

Contents

Introduction

The study of lattice varieties evolved out of the study of varieties in general, which was initiated by Garrett Birkhoff in the 1930's. He derived the first significant results in this subject, and further developments by Alfred Tarski and later, for congruence distributive varieties, by Bjarni Jónsson, laid the groundwork for many of the results about lattice varieties. During the same period, investigations in projective geometry and modular lattices, by Richard Dedekind, John von Neumann, Garrett Birkhoff, George Grätzer, Bjarni Jónsson and others, generated a wealth of information about these structures, which was used by Kirby Baker and Rudolf Wille to obtain some structural results about the lattice of all modular subvarieties. Nonmodular varieties were considered by Ralph McKenzie, and a paper of his published in 1972 stimulated a lot of research in this direction.

Since then the efforts of many people have advanced the subject of lattice varieties in several directions, and many interesting results have been obtained. The purpose of this book is to present a selection of these results in a (more or less) self-contained framework and uniform notation.

In Chapter 1 we recall some preliminary results from the general study of varieties of algebras, and some basic results about congruences on lattices. This chapter also serves to introduce most of the notation which we use subsequently.

Chapter 2 contains some general results about the structure of the lattice Λ of all lattice subvarieties and about the important concept of "splitting". We present several characterizations of splitting lattices and Alan Day's result that splitting lattices generate all lattices. These results are applied in Chapter 4 and 6.

Chapters 3 – 6 each begin with an introduction in which we mention the important results that fall under the heading of the chapter.

Chapter 3 then proceeds with a review of projective spaces and the coordinatization of (complemented) modular lattices. These concepts are used to prove the result of Ralph Freese, that the finite modular lattices do not generate all modular lattices. In the second part of the chapter we give some structural results about covering relations between modular varieties.

In Chapter 4 we concentrate on nonmodular varieties. A characterization of semidistributive varieties is followed by several technical lemmas which lead up to an essentially complete description of the "almost distributive" part of Λ. We derive the result of Bjarni Jónsson and Ivan Rival, that the smallest nonmodular variety has exactly 16 covers, and conclude the chapter with results of Henry Rose about covering chains of join-irreducible semidistributive varieties.

Chapter 5 is concerned with the question which varieties are finitely based. A proof of Kirby Baker's finite basis theorem is followed by an example of a nonfinitely based variety, and a discussion about when the join of two finitely based varieties is again finitely based.

In Chapter 6 we study amalgamation in lattice varieties, and the amalgamation property. The first half of the chapter contains a characterization of the amalgamation class of certain congruence distributive varieties, and in the remaining part we prove that there are only three lattice varieties that have the amalgamation property.

By no means can this monograph be regarded as a full account of the subject of lattice varieties. In particular, the concept of a congruence variety (i.e. the lattice variety generated by the congruence lattices of the members of some variety of algebras) is not included, partly to avoid making this monograph too extensive, and partly because it was felt that this notion is somewhat removed from the topic and requires a wider background of universal algebra.

For the basic concepts and facts from lattice theory and universal algebra we refer the reader to the books of George Grätzer [**GLT**], [**UA**] and Peter Crawley and Robert P. Dilworth [**ATL**]. However, we denote the join of two elements a and b in a lattice by $a + b$ (rather than $a \vee b$) and the meet by $a \cdot b$, or simply ab (instead of $a \wedge b$; for the meet of two congruences we use the symbol $\cap$). When using this plus, dot notation, it is traditionally assumed that the meet operation has priority over the join operation, which reduces the apparent complexity of a lattice expression.

As a final remark, when we consider results that are applicable to wider classes of algebras (not only to lattices) then we aim to state and prove them in an appropriate general form.

Chapter 1

Preliminaries

1.1 The Concept of a Variety

Lattice varieties. Let $\mathcal{E}$ be a set of lattice identities (equations), and denote by $\operatorname{Mod} \mathcal{E}$ the class of all lattices that satisfy every identity in $\mathcal{E}$. A class $\mathcal{V}$ of lattices is a *lattice variety* if

$$\mathcal{V} = \operatorname{Mod} \mathcal{E}$$

for some set of lattice identities $\mathcal{E}$. The class of all lattices, which we will denote by $\mathcal{L}$, is of course a lattice variety since $\mathcal{L} = \operatorname{Mod} \emptyset$. We will also frequently encounter the following lattice varieties:

$$
\begin{aligned}
\mathcal{T} &= \operatorname{Mod}\{x = y\} && - && \text{all trivial lattices,} \\
\mathcal{D} &= \operatorname{Mod}\{xy + xz = x(y + z)\} && - && \text{all distributive lattices,} \\
\mathcal{M} &= \operatorname{Mod}\{xy + xz = x(y + xz)\} && - && \text{all modular lattices.}
\end{aligned}
$$

Let $\mathcal{J}$ be the (countable) set of all lattice identities. For any class $\mathcal{K}$ of lattices, we denote by $\operatorname{Id} \mathcal{K}$ the set of all identities which hold in every member of $\mathcal{K}$. A set of identities $\mathcal{E} \subseteq \mathcal{J}$ is said to be *closed* if

$$\mathcal{E} = \operatorname{Id} \mathcal{K}$$

for some class of lattices $\mathcal{K}$. It is easy to see that for any lattice variety $\mathcal{V}$, and for any closed set of identities $\mathcal{E}$,

$$\mathcal{V} = \operatorname{Mod} \operatorname{Id} \mathcal{V} \quad \text{and} \quad \mathcal{E} = \operatorname{Id} \operatorname{Mod} \mathcal{E},$$

whence there is a bijection between the collection of all lattice varieties, denoted by Λ, and the set of all closed subsets of $\mathcal{J}$. Thus Λ is a set, although its members are proper classes.

Λ is partially ordered by inclusion, and for any collection $\{\mathcal{V}_i : i \in I\}$ of lattice varieties

$$\bigcap_{i \in I} \mathcal{V}_i = \operatorname{Mod} \bigcup_{i \in I} \operatorname{Id} \mathcal{V}_i$$

is again a lattice variety, which implies that Λ is closed under arbitrary intersections. Since Λ also has a largest element, namely $\mathcal{L}$, we conclude that Λ is a complete lattice with intersection as the meet operation. For any class of lattices $\mathcal{K}$,

$$\mathcal{K}^{\mathcal{V}} = \operatorname{Mod} \operatorname{Id} \mathcal{K} = \bigcap \{\mathcal{V} \in \Lambda : \mathcal{K} \subseteq \mathcal{V}\}$$

is the smallest variety containing $\mathcal{K}$, and we call it *the variety generated by $\mathcal{K}$*. Now the join of a collection of lattices varieties is the variety generated by their union. We discuss the lattice Λ in more detail in Section 2.1.

Varieties of algebras. Many of the results about lattice varieties are valid for varieties of other types of algebras, which are defined in a completely analogous way. When we consider a class $\mathcal{K}$ of algebras, then the members of $\mathcal{K}$ are all assumed to be algebras of the same type with only finitary operations. We denote by

$\mathbf{H}\mathcal{K}$ — the class of all *homomorphic images* of members of $\mathcal{K}$

$\mathbf{S}\mathcal{K}$ — the class of all *subalgebras* of members of $\mathcal{K}$

$\mathbf{P}\mathcal{K}$ — the class of all *direct products* of members of $\mathcal{K}$

$\mathbf{P}_S\mathcal{K}$ — the class of all *subdirect products* of members of $\mathcal{K}$.

(Recall that an algebra A is a subdirect product of algebras A_i ($i \in I$) if there is an embedding f from A into the direct product $\bigtimes_{i \in I} A_i$ such that f followed by the ith projection π_i maps A *onto* A_i for each $i \in I$.)

The first significant results in the general study of varieties are due to Birkhoff [35], who showed that varieties are precisely those classes of algebras that are closed under the formation of homomorphic images, subalgebras and direct products, i.e.

$\mathcal{V}$ is a variety if and only if $\quad \mathbf{H}\mathcal{V} = \mathbf{S}\mathcal{V} = \mathbf{P}\mathcal{V} = \mathcal{V}.$

Tarski [46] then put this result in the form

$$\mathcal{K}^{\mathcal{V}} = \mathbf{HSP}\mathcal{K}$$

and later Kogalovskiĭ [65] showed that

$$\mathcal{K}^{\mathcal{V}} = \mathbf{HP}_S\mathcal{K}$$

for any class of algebras $\mathcal{K}$.

1.2   Congruences and Free Algebras

Congruences of algebras. Let A be an algebra, and let $\mathrm{Con}(A)$ be the lattice of all congruences on A. For $a, b \in A$ and $\theta \in \mathrm{Con}(A)$ we denote by:

a/θ — the *congruence class* of a modulo θ

A/θ — the *quotient algebra* of A modulo θ

$\mathbf{0}, \mathbf{1}$ — the *zero* and *unit* of $\mathrm{Con}(A)$

$\mathrm{con}(a, b)$ — the *principal congruence* generated by (a, b)

(i.e. the smallest congruence that identifies a and b).

$\mathrm{Con}(A)$ is of course an algebraic (= compactly generated) lattice with the finite joins of principal congruences as compact elements. (Recall that a lattice element c is *compact* if whenever c is below the join of set of elements C then c is below the join of a finite subset of C. A lattice is *algebraic* if it is complete and every element is a join of compact elements.)

For later reference we recall here a description of the join operation in $\text{Con}(A)$.

LEMMA 1.1 *Let A be an algebra, $x, y \in A$ and $C \subseteq \text{Con}(A)$. Then*

$$x(\textstyle\sum C)y \quad \text{if and only if} \quad x = z_0 \psi_1 z_1 \psi_2 z_2 \cdots z_{n-1} \psi_n z_n = y$$

for some $n \in \omega$, $\psi_1, \psi_2, \ldots, \psi_n \in C$ and $z_0, z_1, \ldots, z_n \in A$.

The connection between congruences and homomorphisms is exhibited by the *homomorphism theorem*: for any homomorphism $h : A \to B$ the image $h(A)$ is isomorphic to $A/\ker h$, where

$$\ker h = \{(a, a') \in A^2 : h(a) = h(a')\} \in \text{Con}(A).$$

We will also make use of the *second isomorphism theorem* which states that for (fixed) $\theta \in \text{Con}(A)$ and any $\phi \in \text{Con}(A)$ containing θ there exists a congruence $(\phi/\theta) \in \text{Con}(A/\theta)$, defined by

$$a/\theta \, (\phi/\theta) \, b/\theta \quad \text{if and only if} \quad a\phi b,$$

such that $(A/\theta)/(\phi/\theta)$ is isomorphic to A/ϕ. Furthermore the map $\phi \mapsto \phi/\theta$ defines an isomorphism from the principal filter $[\theta) = \{\phi \in \text{Con}(A) : \theta \leq \phi\}$ to $\text{Con}(A/\theta)$.

The homomorphism theorem implies that an algebra A is a subdirect product of quotient algebras A/θ_i ($\theta_i \in \text{Con}(A)$) if and only if the meet (intersection) of the θ_i is the **0** of $\text{Con}(A)$. An algebra A is *subdirectly irreducible* if and only if A satisfies anyone of the following equivalent conditions:

  (i) whenever A is a subdirect product of algebras A_i ($i \in I$) then A is isomorphic to one of the factors A_i;

 (ii) the **0** of $\text{Con}(A)$ is completely meet irreducible;

(iii) there exist $a, b \in A$ such that $\text{con}(a, b)$ is the smallest non-**0** element of $\text{Con}(A)$.

A is said to be *finitely subdirectly irreducible* if whenever A is a subdirect product of finitely many algebras $A_1, \ldots, A_n$ then A is isomorphic to one of the A_i ($1 \leq i \leq n$), or equivalently if the **0** of $\text{Con}(A)$ is meet irreducible.

For any variety $\mathcal{V}$ of algebras we denote by

$\mathcal{V}_{SI}$ — the class of all subdirectly irreducible members of $\mathcal{V}$
$\mathcal{V}_{FSI}$ — the class of all finitely subdirectly irreducible members of $\mathcal{V}$.

We can now state Birkhoff's [44] subdirect representation theorem: *Every algebra is a subdirect product of its subdirectly irreducible homomorphic images.*

This can be deduced from the following result concerning decompositions in algebraic lattices:

THEOREM 1.2 ([ATL] p.43). *In an algebraic lattice every element is the meet of completely meet irreducible elements.*

Relatively free algebras. Let $\mathcal{K}$ be a class of algebras, and let F be an algebra generated by a set $X \subseteq F$. We say that F is $\mathcal{K}$-*freely generated* by the set X if any map f from the set of generators X to any $A \in \mathcal{K}$ extends to a homomorphism $\overline{f} : F \to A$. If, in addition, $F \in \mathcal{K}$, then F is called the $\mathcal{K}$- *free algebra on* $\alpha = |X|$ *generators* or the α-*generated free algebra over* $\mathcal{K}$, and is denoted by $F_{\mathcal{K}}(X)$ or $F_{\mathcal{K}}(\alpha)$. The extension $\overline{f}$ is unique (since X generates F), and it follows that $F_{\mathcal{K}}(\alpha)$ is uniquely determined (up to isomorphism) for each cardinal α.

However, $F_{\mathcal{K}}(\alpha)$ need not exist for every class $\mathcal{K}$ of algebras. Birkhoff [**35**] found a simple method of constructing $\mathcal{K}$-freely generated algebras in general, and from this he could deduce that *for any nontrivial variety* $\mathcal{V}$ *and any cardinal* $\alpha \neq 0$, *the* $\mathcal{V}$-*free algebra on* α *generators exists*. We briefly outline his method below (further details can be found in [**UA**]).

Let $W(X)$ be the *word algebra* over the set X, i.e. $W(X)$ is the set of all terms (= polynomials or words) in the language of the algebras in $\mathcal{K}$, with variables taken from the set X, and with the operations defined on $W(X)$ in the obvious way (e.g. for lattices the join of two terms $p, q \in W(X)$ is the term $(p + q) \in W(X)$). It is easy to check that, for any class $\mathcal{K}$ of algebras, $W(X)$ is $\mathcal{K}$-freely generated by the set X. Other $\mathcal{K}$-freely generated algebras can be constructed as quotient algebras of $W(X)$ in the following way: Let

$$\begin{aligned} \theta_{\mathcal{K}} \ &= \ \textstyle\prod \{\ker h \mid h : W(X) \to A \text{ is a homomorphism, } A \in \mathcal{K}\} \\ &(\ = \ \textstyle\sum \{\operatorname{con}(p, q) \mid p, q \in W(X) \text{ and } p = q \in \operatorname{Id}\mathcal{K}\}). \end{aligned}$$

We claim that $F = W(X)/\theta_{\mathcal{K}}$ is $\mathcal{K}$-freely generated by the set $\overline{X} = \{x/\theta_{\mathcal{K}} : x \in X\}$. Indeed, given a map $f : \overline{X} \to A \in \mathcal{K}$, define $f' : X \to A$ by $f'(x) = f(x/\theta_{\mathcal{K}})$, then f' extends to a homomorphism $\overline{f}' : F \to A$. If $\mathcal{K}$ contains a nontrivial algebra, then $|\overline{X}| = |X|$ (if not, then $\theta_{\mathcal{K}}$ identifies all of $W(X)$, hence F is trivial), and by construction F is a subdirect product of the algebras $W(X)/\ker h$, which are all members of $S\mathcal{K}$. Consequently, if $\mathcal{K}$ is closed under the formation of subdirect products, then $F \in \mathcal{K}$, whence Birkhoff's result follows.

If an identity holds in every member of a variety $\mathcal{V}$, then it must hold in $F_{\mathcal{V}}(n)$ for each $n \in \omega$, and if an identity fails in some member of $\mathcal{V}$, then it must fail in some finitely generated algebra (since an identity has only finitely many variables), and hence it fails in $F_{\mathcal{V}}(n)$ for some $n \in \omega$. Thus

$$\mathcal{V} = \{F_{\mathcal{V}}(n) : n \in \omega\}^{\mathcal{V}},$$

and it now follows from Birkhoff's subdirect representation theorem that every variety is generated by its finitely generated subdirectly irreducible members. In fact, a similar argument to the one above shows that

$$\mathcal{V} = \{F_{\mathcal{V}}(\omega)\}^{\mathcal{V}},$$

whence every variety is generated by a single (countably generated) algebra. To obtain an interesting notion of a *finitely generated* variety, we define this to be a variety that is generated by finitely many *finite* algebras, or equivalently, by a finite algebra (the product of the former algebras).

1.3 Congruence Distributivity

An algebra A is said to be *congruence distributive* if the lattice $\text{Con}(A)$ is distributive. A variety $\mathcal{V}$ of algebras is *congruence distributive* if every member of $\mathcal{V}$ is congruence distributive.

Congruence distributive algebras have factorable congruences. What is interesting about algebras in a congruence distributive variety is that they satisfy certain conditions which do not hold in general. The most important ones are described by Jónsson's Lemma (1.4) and its corollaries, but we first point out a result which follows directly from the definition of congruence distributivity.

LEMMA 1.3 *Suppose A is the product of two algebras A_1, A_2 and $\text{Con}(A)$ is distributive. Then $\text{Con}(A)$ is isomorphic to $\text{Con}(A_1) \times \text{Con}(A_2)$ (i.e. congruences on the product of two algebras can be factored into two congruences on the algebras).*

PROOF. Let $L = \text{Con}(A_1) \times \text{Con}(A_2)$, and for $\theta = (\theta_1, \theta_2) \in L$ define a relation $\overline{\theta}$ on A by

$$(a_1, a_2)\overline{\theta}(b_1, b_2) \quad \text{if and only if} \quad a_1\theta_1 b_1 \text{ and } a_2\theta_2 b_2.$$

One easily checks that $\overline{\theta} \in \text{Con}(A)$, and that if $\psi = (\psi_1, \psi_2) \in L$, then

$$\overline{\theta} \subseteq \overline{\psi} \quad \text{if and only if} \quad \theta_1 \subseteq \psi_1 \text{ and } \theta_2 \subseteq \psi_2.$$

Thus the map $\theta \mapsto \overline{\theta}$ from L to $\text{Con}(A)$ is one-one and order preserving, so it remains to show that it is also onto. Let $\phi \in \text{Con}(A)$, and for $i = 1, 2$ define $\rho_i = \ker \pi_i$, where π_i is the projection from A onto A_i. Clearly $\rho_1 \cap \rho_2 = \mathbf{0}$ in $\text{Con}(A)$, hence $\phi = (\phi + \rho_1) \cap (\phi + \rho_2)$ by the distributivity of $\text{Con}(A)$. Observe that for $i = 1, 2$ $\rho_i \subseteq \phi + \rho_i$ and $A/\rho_i \cong A_i$, so from the second isomorphism theorem we obtain $\theta_i \in \text{Con}(A_i)$ such that for $a = (a_1, a_2), b = (b_1, b_2) \in A$

$$a(\phi + \rho_i)b \quad \text{if and only if} \quad a_i\theta_i b_i \qquad (i \in \{1, 2\}).$$

Therefore $a\phi b$ iff $a(\phi + \rho_1)b$ and $a(\phi + \rho_2)b$ iff $a_1\theta_1 b_1$ and $a_2\theta_2 b_2$. Letting $\theta = (\theta_1, \theta_2) \in L$, we see that $\phi = \overline{\theta}$. □

A short review of filters, ideals and ultraproducts. Let L be a lattice and F a filter in L (i.e. F is a sublattice of L, and for all $y \in L$, if $y \geq x \in F$ then $y \in F$). A filter F is *proper* if $F \neq L$, it is *principal* if $F = [a) = \{x \in L : a \leq x\}$ for some $a \in L$, and F is *prime* if $x + y \in F$ implies $x \in F$ or $y \in F$ for all $x, y \in L$. An *ultrafilter* is a maximal proper filter (maximal with respect to inclusion). In a distributive lattice the notions of a proper prime filter and an ultrafilter coincide.

The notion of a (proper / principal / prime) ideal is defined dually, and principal ideals are denoted by $(a]$. Let $\mathcal{I}L$ be the collection of all ideals in L. Then $\mathcal{I}L$ is closed under arbitrary intersections and has a largest element, hence it is a lattice, partially ordered by inclusion. Dually, the collection of all filters of L, denoted by $\mathcal{F}L$, is also a lattice. The order on $\mathcal{F}L$ is *reverse inclusion* i.e. $F \leq G$ iff $F \supseteq G$. With these definitions it is not difficult to see that the map $x \mapsto [x)$ $(x \mapsto (x])$ is an embedding of L into $\mathcal{F}L$ $(\mathcal{I}L)$, and that $\mathcal{F}L$ and $\mathcal{I}L$ satisfy the same identities as L. It follows that whenever L is a member of some lattice variety then so are $\mathcal{F}L$ and $\mathcal{I}L$.

For an arbitrary set I we say that $\mathcal{F}$ *is a filter over* I if $\mathcal{F}$ is a filter in the *powerset lattice* $\mathcal{P}I$ (= the collection of all subsets of I ordered by inclusion). Since $\mathcal{P}I$ is distributive, a filter $\mathcal{U}$ is an ultrafilter if and only if $\mathcal{U}$ is a proper prime filter, and this is equivalent to the condition that whenever I is partitioned into finitely many disjoint blocks, then $\mathcal{U}$ contains exactly one of these blocks.

Let $A = \bigtimes_{i \in I} A_i$ be a direct product of a family of algebras $\{A_i : i \in I\}$. If $\mathcal{F}$ is a filter over I (i.e. a filter in the powerset lattice $\mathcal{P}I$), then we can define a congruence $\phi_{\mathcal{F}}$ on A by

$$a\phi_{\mathcal{F}}b \quad \text{if and only if} \quad \{i \in I : a_i = b_i\} \in \mathcal{F}$$

where a_i is the ith coordinate of a.

If A is a direct product of algebras A_i $(i \in I)$ and $\mathcal{U}$ is an ultrafilter over I, then the quotient algebra $A/\phi_{\mathcal{U}}$ is called an *ultraproduct*. For any given class of algebras $\mathcal{K}$, we denote by

$$\mathbf{P}_U\mathcal{K} \text{ — the class of all ultraproducts of members of } \mathcal{K}.$$

Jónsson's Lemma. We are now ready to state and prove this remarkable result.

LEMMA 1.4 (Jónsson [**67**]). *Suppose B is a congruence distributive subalgebra of a direct product $A = \bigtimes_{i \in I} A_i$, and $\theta \in \mathrm{Con}(B)$ is meet irreducible. Then there exists an ultrafilter $\mathcal{U}$ over I such that $\phi_{\mathcal{U}}|B \subseteq \theta$.*

PROOF. We will denote by $[J]$ the principal filter generated by a subset J of I, and to simplify the notation we set $\psi_J = \phi_{[J]}|B$. Clearly $a\psi_J b$ if and only if $J \subseteq \{i \in I : a_i = b_i\}$, for $a, b \in B \subseteq A$. If $\theta = 1 \in \mathrm{Con}(B)$ then any ultrafilter over I will do. So assume $\theta < 1$ and let $\mathcal{C}$ be the collection of all subsets J of I such that $\psi_J \subseteq \theta$. We claim that $\mathcal{C}$ has the following properties:

(i) $I \in \mathcal{C}, \emptyset \notin \mathcal{C}$;

(ii) $K \supseteq J \in \mathcal{C}$ implies $K \in \mathcal{C}$;

(iii) $J \cup K \in \mathcal{C}$ implies $J \in \mathcal{C}$ or $K \in \mathcal{C}$.

(i) and (ii) hold because $\psi_I, \psi_\emptyset$ are the zero and unit of $\mathrm{Con}(B)$, and $K \supseteq J$ implies $\psi_K \le \psi_J$. To prove (iii), observe that $J \cup K \in \mathcal{C}$ implies

$$\theta = \theta + \psi_{J \cup K} = \theta + (\psi_J \cap \psi_K) = (\theta + \psi_J) \cap (\theta + \psi_K)$$

by the distributivity of $\mathrm{Con}(B)$, and since θ is assumed to be meet irreducible, it follows that $\theta = \theta + \psi_J$ or $\theta = \theta + \psi_K$, whence $J \in \mathcal{C}$ or $K \in \mathcal{C}$.

$\mathcal{C}$ itself need not be a filter, but using Zorn's Lemma we can choose a filter $\mathcal{U}$ over I, maximal with respect to the property $\mathcal{U} \subseteq \mathcal{C}$. It is easy to see that

$$\phi_{\mathcal{U}}|B = \bigcup_{J \in \mathcal{U}} \psi_J \subseteq \theta$$

so it remains to show that $\mathcal{U}$ is an ultrafilter. Suppose the contrary. Then there exists a set $H \subseteq I$ such that neither H nor $I - H$ belong to $\mathcal{U}$. If $H \cap J \in \mathcal{C}$ for all $J \in \mathcal{U}$, then by (i) and (ii) $\mathcal{U} \cup \{H\}$ would generate a filter contained in $\mathcal{C}$, contradicting the maximality of

$\mathcal{U}$, and similarly for $I - H$. Hence we can find $J, K \in \mathcal{U}$ such that $H \cap J, (I - H) \cap K \notin \mathcal{C}$. But $J \cap K \in \mathcal{U}$, and

$$J \cap K = (H \cap J \cap K) \cup ((I - H) \cap J \cap K)$$

which contradicts (iii). This completes the proof. $\square$

COROLLARY 1.5 (Jónsson [**67**]). *Let $\mathcal{K}$ be a class of algebras such that $\mathcal{V} = \mathcal{K}^{\mathcal{V}}$ is congruence distributive. Then*

(i) $\mathcal{V}_{SI} \subseteq \mathcal{V}_{FSI} \subseteq \mathbf{HSP}_U \mathcal{K}$;

(ii) $\mathcal{V} = \mathbf{P}_S \mathbf{HSP}_U \mathcal{K}$.

PROOF. (i) We always have $\mathcal{V}_{SI} \subseteq \mathcal{V}_{FSI}$. Since $\mathcal{V} = \mathbf{HSP}\mathcal{K}$, every algebra in $\mathcal{V}$ is isomorphic to a quotient algebra B/θ, where B is a subalgebra of a direct product $A = \bigtimes_{i \in I} A_i$ and $\{A_i : i \in I\} \subseteq \mathcal{K}$. If B/θ is finitely subdirectly irreducible, then θ is meet irreducible, hence by the preceding lemma there exists an ultrafilter $\mathcal{U}$ over I such that $\phi = \phi_{\mathcal{U}} | B \subseteq \theta$. Thus B/θ is a homomorphic image of B/ϕ, which is isomorphic to a subalgebra of the ultraproduct $A/\phi_{\mathcal{U}}$. (ii) follows from (i) and Birkhoff's subdirect representation theorem. $\square$

To exhibit the full strength of Jónsson's Lemma for finitely generated varieties, we need the following result:

LEMMA 1.6 (Frayne, Morel and Scott [**62**]). *If $\mathcal{K}$ is a finite set of finite algebras, then every ultraproduct of members of $\mathcal{K}$ is isomorphic to a member of $\mathcal{K}$.*

PROOF. Let $A = \bigtimes_{i \in I} A_i$ be a direct product of members of $\mathcal{K}$, and define an equivalence relation $\sim$ on I by $i \sim j$ if and only if $A_i \cong A_j$. Since $\mathcal{K}$ is finite, $\sim$ partitions I into finitely many blocks $I_0, I_1, \ldots, I_n$. If $\mathcal{U}$ is an ultrafilter over I, then $\mathcal{U}$ contains exactly one of these blocks, say $J = I_k$. Let $\overline{\mathcal{U}} = \{U \cap J : U \in \mathcal{U}\}$ be the ultrafilter over J induced by $\mathcal{U}$, and let $\overline{A} = \bigtimes_{j \in J} A_j$. We claim that

$$A/\phi_{\mathcal{U}} \cong \overline{A}/\phi_{\overline{\mathcal{U}}} \cong B$$

where B is an algebra isomorphic to each of the A_j ($j \in J$), and hence to a member of $\mathcal{K}$ as required.

Consider the epimorphism $h : A \to \overline{A}/\phi_{\overline{\mathcal{U}}}$ given by $h(a) = \overline{a}/\phi_{\overline{\mathcal{U}}}$, where $\overline{a}$ is the restriction of a to J. We have

$$a \ker h \ b \quad \text{iff} \quad \overline{a}\phi_{\overline{\mathcal{U}}}\overline{b} \quad \text{iff} \quad \{j \in J : a_j = b_j\} \in \overline{\mathcal{U}}$$
$$\text{iff} \quad \{i \in I : a_i = b_i\} \in \mathcal{U} \quad \text{iff} \quad a\phi_{\mathcal{U}}b$$

so $\ker h = \phi_{\mathcal{U}}$, whence the first isomorphism follows. To establish the second isomorphism, observe that $\overline{A} \cong B^J$, and therefore $\overline{A}/\phi_{\overline{\mathcal{U}}}$ is isomorphic to an ultrapower $B^J/\phi_{\overline{\mathcal{U}}}$ over the finite algebra B. In this case the canonical embedding $B \hookrightarrow B^J/\phi_{\overline{\mathcal{U}}}$, given by $b \mapsto \hat{b}/\phi_{\overline{\mathcal{U}}}$ ($\hat{b} = (b, b, \ldots) \in B^J$), is always onto (hence an isomorphism) because for each $c \in B^J$ we can partition J into finitely many blocks $J_{c,b} = \{j \in J : c_j = b\}$ (one for each $b \in B$), and then one of these blocks, say $J_{c,b'}$ must be in $\overline{\mathcal{U}}$, hence $b' \mapsto c/\phi_{\overline{\mathcal{U}}}$. $\square$

COROLLARY 1.7 (Jónsson [**67**]). *Let* $\mathcal{K}$ *be a finite set of finite algebras such that* $\mathcal{V} = \mathcal{K}^{\mathcal{V}}$
is congruence distributive. Then

(i) $\mathcal{V}_{SI} \subseteq \mathcal{V}_{FSI} \subseteq \mathbf{HS}\mathcal{K}$;

(ii) $\mathcal{V}$ *has up to isomorphism only finitely many subdirectly irreducible members, each*
one finite;

(iii) $\mathcal{V}$ *has only finitely many subvarieties;*

(iv) *if* $A, B \in \mathcal{V}_{SI}$ *are nonisomorphic and* $|A| \leq |B|$, *then there is an identity that holds*
in A but not in B.

PROOF. (i) follows immediately from 1.5 and 1.6, and (ii) follows from (i), since $\mathbf{HS}\mathcal{K}$
has only finitely many non isomorphic members. (iii) holds because every subvariety is
determined by its subdirectly irreducible members, and (iv) follows from the observation
that if both A and B are finite, nonisomorphic and $|A| \leq |B|$, then $B \notin \mathbf{HS}\{A\}$, which
implies $B \notin \{A\}^{\mathcal{V}}$ by part (i).   □

Lattices are congruence distributive algebras. This is one of the most important
results about lattices, since it means that we can apply Jónsson's Lemma to lattices. We
first give a direct proof of this result.

THEOREM 1.8 (Funayama and Nakayama [**42**]). *For any lattice L*, $\mathrm{Con}(L)$ *is distributive.*

PROOF. Let $\theta, \psi, \phi \in \mathrm{Con}(L)$ and observe that the inclusion $(\theta \cap \psi) + (\theta \cap \phi) \subseteq \theta \cap (\psi + \phi)$
holds in any lattice. So suppose for some $x, y \in L$ the congruence $\theta \cap (\psi + \phi)$ identifies x
and y. We have to show that $(\theta \cap \psi) + (\theta \cap \phi)$ identifies x and y. By assumption $x\theta y$ and
$x(\psi+\phi)y$, hence by Lemma 1.1

$$x = z_0 \psi z_1 \phi z_2 \psi z_3 \phi z_4 \cdots z_n = y$$

for some $z_0, z_1, \ldots, z_n \in L$. If we can replace the elements z_i by z_i', which all belong to the
same θ-class as x and y, then

$$x = z_0'(\theta \cap \psi) z_1'(\theta \cap \phi) z_2'(\theta \cap \psi) z_3'(\theta \cap \phi) z_4' \cdots z_n' = y,$$

whence $x(\theta \cap \psi) + (\theta \cap \phi)y$ follows. One way of making this replacement is by taking
$z_i' = xz_i + yz_i + xy$ (the median polynomial), then any congruence which identifies z_i with
z_{i+1}, also identifies z_i' with z_{i+1}', and (since L is a lattice) $xy \leq z_i' \leq x+y$, whence $z_i' \in x/\theta$
for all $i = 0, 1, \ldots, n$.   □

Consequently every lattice variety is congruence distributive. Notice that the proof
appeals to the lattice properties of L only in the last few lines.

Jónsson polynomials. The next theorem is a generalization of the above result.

THEOREM 1.9 (Jónsson [**67**]). *A variety* $\mathcal{V}$ *of algebras is congruence distributive if and*
only if for some positive integer n there exist ternary polynomials $t_0, t_1, \ldots, t_n$ *such that*
for $i = 0, 1, \ldots, n$, *the following identities hold in* $\mathcal{V}$:

$$
\begin{aligned}
&t_0(x,y,z) = x, \qquad t_n(x,y,z) = z, \qquad t_i(x,y,x) = x \\
(*)\quad &t_i(x,x,z) = t_{i+1}(x,x,z) \qquad \text{for } i \text{ even} \\
&t_i(x,z,z) = t_{i+1}(x,z,z) \qquad \text{for } i \text{ odd.}
\end{aligned}
$$

PROOF. Suppose $\mathcal{V}$ is congruence distributive and consider the algebra $F_{\mathcal{V}}(\{a,b,c\})$. Define $\theta = \mathrm{con}(a,c)$, $\phi = \mathrm{con}(a,b)$ and $\psi = \mathrm{con}(b,c)$. Then $(a,c) \in \theta \cap (\phi + \psi) = (\theta \cap \phi) + (\theta \cap \psi)$. By Lemma 1.1 there exist $d_0, d_1, \ldots, d_n \in A$ such that

$$a = d_0 \, \theta \cap \phi \, d_1 \, \theta \cap \psi \, d_2 \, \theta \cap \phi \, d_3 \, \theta \cap \psi \cdots d_n = c.$$

Each d_i is of the form $d_i = t_i(a,b,c)$ for some ternary polynomial t_i, and it remains to show that the identities (∗) are satisfied for $x = a$, $y = b$ and $z = c$ in $F_{\mathcal{V}}(\{a,b,c\})$, since then they must hold in every member of $\mathcal{V}$. The first two identities follow from the fact that $d_0 = a$ and $d_n = c$. For the third identity let $h' : F_{\mathcal{V}}(\{a,b,c\}) \to F_{\mathcal{V}}(\{a,b\})$ be the homomorphism induced by the map $a, c \mapsto a$, $b \mapsto b$. Then $h(a) = h(c)$ implies $\theta \subseteq \ker h$ and since each $d_i = t_i(a,b,c)$ is θ-related to a we have

$$a = h(a) = h(t_i(a,b,c)) = t_i(a,b,a).$$

Now suppose i is even and consider $h : F_{\mathcal{V}}(\{a,b,c\}) \to F_{\mathcal{V}}(\{a,c\})$ induced by the map $a, b \mapsto a$, $c \mapsto c$. Then $\phi \subseteq \ker h$, and since

$$t_i(a,b,c) \, \phi \, t_{i+1}(a,b,c)$$

it follows that

$$t_i(a,a,c) = h(t_i(a,b,c)) = h(t_{i+1}(a,b,c)) = t_{i+1}(a,a,c).$$

The proof for odd i is similar.

Now assume the identities (∗) hold in $\mathcal{V}$ for some ternary polynomials $t_0, t_1, \ldots, t_n$, let $A \in \mathcal{V}$ and $\theta, \phi, \psi \in \mathrm{Con}(A)$. To prove that $\mathcal{V}$ is congruence distributive it suffices to show that $\theta \cap (\phi + \psi) \subseteq (\theta \cap \phi) + (\theta \cap \psi)$. Let $(a,c) \in \theta \cap (\phi + \psi)$. Then $(a,c) \in \theta$ and by Lemma 1.1 there exist $b_0, b_1, \ldots, b_m \in A$ such that

$$a = b_0 \, \phi \, b_1 \, \psi \, b_2 \, \phi \, b_3 \, \psi \cdots b_m = c.$$

So for each $i = 0, 1, \ldots, n$ we have

$$t_i(a,b_0,c) \, \phi \, t_i(a,b_1,c) \, \psi \, t_i(a,b_2,c) \, \phi \cdots t_i(a,b_m,c).$$

The identity $t_i(x,y,x) = x$ together with $(a,c) \in \theta$ implies that the elements $t_i(a,b_j,c) \in a/\theta$ whence

$$t_i(a,b_0,c) \, \theta \cap \phi \, t_i(a,b_1,c) \, \theta \cap \psi \, t_i(a,b_2,c) \, \theta \cap \phi \cdots t_i(a,b_m,c).$$

It follows that $t_i(a,a,c) \, (\theta \cap \phi) + (\theta \cap \psi) \, t_i(a,c,c)$ holds for each $i = 0, 1, \ldots, n$ and the remaining identities now give $(a,c) \in (\theta \cap \phi) + (\theta \cap \psi)$. $\square$

The polynomials $t_0, t_1, \ldots, t_n$ are known as Jónsson polynomials, and will be of use in Chapter 5. Here we just note that for lattices we can deduce Theorem 1.8 from Theorem 1.9 if we take $n = 2$, $t_0(x,y,z) = x$, $t_1(x,y,z) = xy + zy + zx$ (the median polynomial) and $t_2(x,y,z) = z$.

1.4   Congruences on Lattices

Prime quotients and unique maximal congruences. Let L be a lattice and $u, v \in L$ with $v \leq u$. By a *quotient* u/v (alternatively *interval* $[v, u]$) we mean the sublattice $\{x \in L : v \leq x \leq u\}$. We say that u/v is *nontrivial* if $u > v$, and *prime* if u covers v (i.e. $u/v = \{u, v\}$, in symbols: $u \succ v$). If L is subdirectly irreducible and $\mathrm{con}(u, v)$ is the smallest non-0 congruence of L, then u/v is said to be a *critical* quotient.

LEMMA 1.10 *If u/v is a prime quotient of L, then there exists a unique maximal congruence θ that does not identify u and v.*

PROOF. Let $\mathcal{C} \subseteq \mathrm{Con}(L)$ be the set of all congruences of L that do not identify u and v. Take $\theta = \sum \mathcal{C}$, and suppose θ identifies u and v. By Lemma 1.1 we can find $\psi_1, \psi_2, \ldots, \psi_n \in \mathcal{C}$ and $z_0, z_1, \ldots, z_n \in L$ such that

$$u = z_0 \psi_1 z_1 \psi_2 z_2 \cdots z_{n-1} \psi_n z_n = v.$$

Replacing z_i by $z_i' = u z_i + v z_i + v$ we see that $u = z_0' \psi_1 z_1' \psi_2 z_2' \cdots z_{n-1}' \psi_n z_n' = v$ and $v \leq z_i' \leq u$ for all $i = 1, \ldots, n$. Since u/v is assumed to be prime, we must have $z_i' = u$ or $z_i' = v$ for all i, which implies $u \psi_i v$ for some i, a contradiction. Thus $\theta \in \mathcal{C}$, and it is clearly the largest element of $\mathcal{C}$.   □

Weak transpositions. Given two quotients r/s and u/v in L, we say that r/s *transposes weakly up onto* u/v (in symbols $r/s \nearrow_w u/v$) if $r + v = u$ and $s \leq v$. Dually, we say that r/s *transposes weakly down onto* u/v (in symbols $r/s \searrow_w u/v$) if $su = v$ and $r \geq u$. We write $r/s \sim_w u/v$ if r/s transposes weakly up or down onto u/v. The quotient r/s *projects weakly onto* u/v in n *steps* if there exists a sequence of quotients x_i/y_i in L such that

$$r/s = x_0/y_0 \sim_w x_1/y_1 \sim_w \ldots \sim_w x_n/y_n = u/v.$$

Note that the symbols $\nearrow_w$, $\searrow_w$ and $\sim_w$ define nonsymmetric binary relations on the set of quotients of a lattice. Some authors (in particular [GLT], [ATL] and Rose [84]) define weak transpositions in terms of the *inverses* of the above relations, but denote these inverse relations by the same symbols. Usually the phrase "transposes weakly into" (rather than "onto") is used to distinguish the two definitions.

The usefulness of weak transpositions lies in the fact that they can be used to characterize principal congruences in arbitrary lattices.

LEMMA 1.11 (Dilworth [50]). *Let r/s and u/v be quotients in a lattice L. Then $\mathrm{con}(r, s)$ identifies u and v if and only if for some finite chain $u = t_0 > t_1 > \ldots > t_m = v$, the quotient r/s projects weakly onto t_i/t_{i+1} (all $i = 0, 1, \ldots, m - 1$).*

Notice that if u/v is a prime critical quotient of a subdirectly irreducible lattice L, then by the above lemma every nontrivial quotient of L projects weakly onto u/v.

Bijective transpositions and modularity. We say that r/s *transposes up onto* u/v (in symbols $r/s \nearrow u/v$) or equivalently u/v *transposes down onto* r/s (in symbols $u/v \searrow r/s$) if $r + v = u$ and $rv = s$. We write $r/s \sim u/v$ if either $r/s \nearrow u/v$ or $r/s \searrow u/v$. Note that $\sim$ is a symmetric relation, and that

$$r/s \sim_w u/v \text{ and } u/v \sim_w r/s \quad \text{if and only if} \quad r/s \sim u/v.$$

Suppose $r/s \nearrow u/v$ and, in addition, for every $t \in r/s$ and $t' \in u/v$ we have

$$t = (t + v)r \qquad \text{and} \qquad t' = t'r + v.$$

Then the map $t \mapsto t + v$ is an isomorphism from r/s to u/v, and in this case we say that r/s *transposes bijectively up onto* u/v (in symbols $r/s \nearrow_\beta u/v$), or equivalently u/v *transposes bijectively down onto* r/s (in symbols $u/v \searrow_\beta r/s$). In a modular lattice every transpose is bijective, since $t \le r$ and modularity imply

$$(t + v)r = t + vr = t + s = t$$

and similarly $t' = t'r + v$. It follows that for any sequence of weak transpositions $x_0/y_0 \sim_w x_1/y_1 \sim_w \ldots \sim_w x_n/y_n$ we can find subquotients x_i'/y_i' of x_i/y_i $(i = 0, 1, \ldots, n-1)$ such that

$$x_0/y_0 \supseteq x_0'/y_0' \sim x_1'/y_1' \sim \ldots \sim x_n/y_n.$$

In this case we say that the two quotients x_0'/y_0' and x_n/y_n are *projective to each other*, and by Lemma 1.11 this concept is clearly sufficient for describing principal congruences in modular lattices.

Congruence lattices of modular lattices. The symbol **2** denotes a two element lattice, and a complemented distributive lattice will be referred to as a Boolean algebra (although we do not include complementation, zero and unit as basic operations). We need the following elementary result about distributive lattices:

LEMMA 1.12 *Let D be a finite distributive lattice. If the largest element of D is a join of atoms, then D is a Boolean algebra.*

PROOF. It suffices to show that D is complemented. Let $a \in D$ and define $\overline{a}$ to be the join of all atoms that are not below a. By assumption $a + \overline{a} = 1_D$ and by distributivity $a\overline{a} = 0_D$, whence $\overline{a}$ is the complement of a. $\qquad\square$

A *chain* C is a linearly ordered subset of a lattice, and if $|C|$ is finite then the *length* of C is defined to be $|C| - 1$. A lattice L is said to be *of length* n if there is a chain in L that has length n and all chains in L are of length $\le n$. Recall the Jordan-Hölder Chain condition ([**GLT**] p.172): if M is a (semi-) modular lattice of finite length then any two maximal chains in M have the same length. In such lattices the length is also referred to as the *dimension* of the lattice.

LEMMA 1.13 *Let M be a modular lattice.*

(i) *If u/v is a prime quotient of M, then $con(u, v)$ is an atom of $Con(M)$.*

(ii) *If M has finite length m, then $Con(M)$ is isomorphic to a Boolean algebra $\mathbf{2}^n$, where $n \le m$.*

PROOF. (i) If $con(u, v) \supseteq con(r, s)$ for some $r \ne s \in M$, then u/v and a prime subquotient of $r+s/rs$ are projective to each other, which implies that $con(u, v) = con(r, s)$. It follows that $con(u, v)$ is an atom.

(ii) Let $z_0 \prec z_1 \prec \ldots \prec z_m$ be a maximal chain in M. Then the principal congruences $con(z_i, z_{i+1})$ $(i = 0, 1, \ldots, m-1)$ are atoms (not necessarily distinct) of $Con(M)$, and since their join collapses the whole of M, the result follows from the distributivity of $\mathrm{Con}(M)$ and the preceding lemma.   □

As a corollary we have that every subdirectly irreducible modular lattice of finite length is *simple* (i.e. $\mathrm{Con}(M) \cong \mathbf{2}$).

Chapter 2

General Results

2.1 The Lattice Λ

Λ is a dually algebraic distributive lattice. In Section 1.1 it was shown that the collection Λ of all lattice subvarieties of $\mathcal{L}$ is a complete lattice, with intersection as meet. A completely analogous argument shows that this result is true in general for the collection of all subvarieties of an arbitrary variety $\mathcal{V}$ of algebras. We denote by

$$\Lambda_\mathcal{V} \text{ — the lattice of all subvarieties of the variety } \mathcal{V}.$$

(If $\mathcal{V} = \mathcal{L}$ then we usually drop the subscript $\mathcal{V}$.)

Call a variety $\mathcal{V}' \in \Lambda_\mathcal{V}$ *finitely based relative to* $\mathcal{V}$ if it can be defined by some finite set of identities together with the set Id $\mathcal{V}$. If $\mathcal{V}$ is finitely based relative to the variety Mod $\emptyset$ (= the class of all algebras of the same type as $\mathcal{V}$) then we may omit the phrase "relative to $\mathcal{V}$".

THEOREM 2.1 *For any variety $\mathcal{V}$ of algebras, $\Lambda_\mathcal{V}$ is a dually algebraic lattice, and the varieties which are finitely based relative to $\mathcal{V}$ are the dually compact elements.*

PROOF. Let $\mathcal{V}'$, $\mathcal{V}_i$ ($i \in I$) be subvarieties of $\mathcal{V}$, and suppose that $\mathcal{V}' \supseteq \bigcap_{i \in I} \mathcal{V}_i = \text{Mod}(\bigcup_{i \in I} \text{Id}\mathcal{V}_i)$. If $\mathcal{V}'$ is finitely based relative to $\mathcal{V}$, then $\mathcal{V}' = \mathcal{V} \cap \text{Mod}\mathcal{E}$ for some finite set $\mathcal{E} \subseteq \text{Id}\mathcal{V}'$. It follows that $\mathcal{E} \subseteq \bigcup_{i \in I} \text{Id}\mathcal{V}_i$, and since $\mathcal{E}$ is finite, it will be included in the union of finitely many $\text{Id}\mathcal{V}_i$. Clearly the finite intersection of the corresponding subvarieties is included in $\mathcal{V}'$, whence $\mathcal{V}'$ is dually compact.

Conversely, suppose $\mathcal{V}'$ is dually compact. We always have

$$(*) \qquad \mathcal{V}' = \text{Mod Id } \mathcal{V}' = \mathcal{V} \cap \bigcap_{\varepsilon \in \text{Id}\mathcal{V}'} \text{Mod}\{\varepsilon\},$$

so by dual compactness $\mathcal{V}' = \mathcal{V} \cap \bigcap_{i=1}^{n} \text{Mod}\{\varepsilon_i\}$ for some finite set $\mathcal{E} = \{\varepsilon_1, \ldots, \varepsilon_n\} \subseteq \text{Id}\mathcal{V}'$. Hence $\mathcal{V}'$ is finitely based relative to $\mathcal{V}$.

Finally $(*)$ implies that every element of $\Lambda_\mathcal{V}$ is a meet of dually compact elements, and so $\Lambda_\mathcal{V}$ is dually algebraic. □

Let $C_\mathcal{V}(\mathcal{V}')$ denote the collection of all varieties in $\Lambda_\mathcal{V}$ that cover $\mathcal{V}'$. We say that $C_\mathcal{V}(\mathcal{V}')$ *strongly covers* $\mathcal{V}'$ if any variety that properly contains $\mathcal{V}'$, contains some member of $C_\mathcal{V}(\mathcal{V}')$.

Recall that a lattice L is *weakly atomic* if every nontrivial quotient of L contains a prime subquotient. An algebraic (or dually algebraic) lattice L is always weakly atomic, since for any nontrivial quotient u/v in L we can find a compact element $c \leq u$, $c \not\leq v$ and using Zorn's Lemma we can choose a maximal element d of the set $\{x \in L : v \leq d < c+v\}$, which then satisfies $v \leq d \prec c + v \leq u$. In particular, if u is compact, then there exists $d \in L$ such that $v \leq d \prec u$.

THEOREM 2.2 *Let* $\mathcal{V}'$ *be a subvariety of a variety* $\mathcal{V}$.

(i) *If* $\mathcal{V}'$ *is finitely based relative to* $\mathcal{V}$ *then* $C_{\mathcal{V}}(\mathcal{V}')$ *strongly covers* $\mathcal{V}'$.

(ii) *If* $C_{\mathcal{V}}(\mathcal{V}')$ *is finite and strongly covers* $\mathcal{V}'$ *then* $\mathcal{V}'$ *is finitely based relative to* $\mathcal{V}$.

PROOF. (i) $\mathcal{V}'$ is dually compact, so by the remark above, any variety which contains $\mathcal{V}'$, contains a variety that covers $\mathcal{V}'$.

(ii) Suppose $C_{\mathcal{V}}(\mathcal{V}') = \{\mathcal{V}_1, \ldots, \mathcal{V}_n\}$ for some $n \in \omega$. Then for each $i = 1, \ldots, n$ there exists an identity $\varepsilon_i \in \mathrm{Id}\,\mathcal{V}'$ such that ε_i fails in $\mathcal{V}_i$. Let $\mathcal{V}'' = \mathcal{V} \cap \mathrm{Mod}\{\varepsilon_i : i = 1, \ldots, n\}$. We claim that $\mathcal{V}' = \mathcal{V}''$.

Since each ε_i holds in $\mathcal{V}'$, we certainly have $\mathcal{V}' \subseteq \mathcal{V}''$. If $\mathcal{V}' \neq \mathcal{V}''$ then the assumption that $C_{\mathcal{V}}(\mathcal{V}')$ strongly covers $\mathcal{V}'$ implies that $\mathcal{V}_i \subseteq \mathcal{V}''$ for some $i \in \{1, \ldots, n\}$. But this is a contradiction since ε_i fails in $\mathcal{V}_i$.   □

We now focus our attention on congruence distributive varieties, since we can then apply Jónsson's Lemma to obtain further results.

THEOREM 2.3 (Jónsson [67]). *Let* $\mathcal{V}$ *be a congruence distributive variety of algebras and let* $\mathcal{V}', \mathcal{V}'' \in \Lambda_{\mathcal{V}}$. *Then*

(i) $(\mathcal{V}' + \mathcal{V}'')_{SI} = \mathcal{V}'_{SI} \cup \mathcal{V}''_{SI}$;

(ii) $\Lambda_{\mathcal{V}}$ *is a distributive lattice;*

(iii) *if* $\mathcal{V}'$ *is finitely generated, then* $\mathcal{V}' + \mathcal{V}''/\mathcal{V}''$ *is a finite quotient in* $\Lambda_{\mathcal{V}}$.

PROOF. (i) We always have $\mathcal{V}'_{SI} \cup \mathcal{V}''_{SI} \subseteq (\mathcal{V}' + \mathcal{V}'')_{SI}$. Conversely, if $A \in (\mathcal{V}' + \mathcal{V}'')_{SI}$ then Jónsson's Lemma implies that $A \in \mathbf{HSP}_U(\mathcal{V}' \cup \mathcal{V}'')$. It is not difficult to see that $\mathbf{HSP}_U(\mathcal{V}' \cup \mathcal{V}'') = \mathbf{HSP}_U\mathcal{V}' \cup \mathbf{HSP}_U\mathcal{V}'' = \mathcal{V}' \cup \mathcal{V}''$, and since A is subdirectly irreducible, we must have $A \in \mathcal{V}'_{SI}$ or $A \in \mathcal{V}''_{SI}$.

(ii) If $\mathcal{V}_1, \mathcal{V}_2, \mathcal{V}_3 \in \Lambda_{\mathcal{V}}$, then (i) implies that every subdirectly irreducible member of $\mathcal{V}_1 \cap (\mathcal{V}_2 + \mathcal{V}_3)$ belongs to either $\mathcal{V}_1 \cap \mathcal{V}_2$ or $\mathcal{V}_1 \cap \mathcal{V}_3$, whence $\mathcal{V}_1 \cap (\mathcal{V}_2 + \mathcal{V}_3) \subseteq (\mathcal{V}_1 \cap \mathcal{V}_2) + (\mathcal{V}_1 \cap \mathcal{V}_3)$. The reverse inclusion is always satisfied.

(iii) By Corollary 1.7(iii), the quotient $\mathcal{V}'/\mathcal{V}' \cap \mathcal{V}''$ is finite, and it transposes bijectively up onto $\mathcal{V}' + \mathcal{V}''/\mathcal{V}''$ since $\Lambda_{\mathcal{V}}$ is distributive.   □

The fact that, for any congruence distributive variety $\mathcal{V}$, the lattice $\Lambda_{\mathcal{V}}$ is dually algebraic and distributive can also be derived from the following more general result, due to B. H. Neumann [62]:

THEOREM 2.4 *For any variety* $\mathcal{V}$ *of algebras,* $\Lambda_{\mathcal{V}}$ *is dually isomorphic to the lattice of all fully invariant congruences on* $F_{\mathcal{V}}(\omega)$.

(A congruence $\theta \in \mathrm{Con}(A)$ is *fully invariant* if $a\theta b$ implies $f(a)\theta f(b)$ for all endomorphisms $f : A \hookrightarrow A$). However, we will not make use of this result.

Some properties of the variety $\mathcal{L}$. For any class $\mathcal{K}$ of algebras, denote by

$$\mathcal{K}_F \quad — \text{ the class of all finite members of } \mathcal{K}.$$

The variety $\mathcal{V} = \mathcal{L}$ of all lattice varieties has the following interesting properties:

(P1) $\mathcal{V}$ is generated by its finite (subdirectly irreducible) members (i.e. $\mathcal{V} = (\mathcal{V}_F)^{\mathcal{V}}$).

(P2) Every member of $\mathcal{V}$ can be embedded in a member of $\mathcal{V}_{SI}$ (i.e. $\mathcal{V} \subseteq S\mathcal{V}_{SI}$).

(P3) Every finite member of $\mathcal{V}$ can be embedded in a finite member of $\mathcal{V}_{SI}$.

That $\mathcal{L}$ satisfies (P1) was proved by Dean [**56**], who showed that if an identity fails in some lattice, then it fails in some finite lattice (see Lemma 2.23). In Section 2.3 we prove an even stronger result, namely that $\mathcal{L}$ is generated by the class of all splitting lattices (which are all finite). (P2) follows from the result of Whitman [**46**] that every lattice can be embedded in a partition lattice, which is simple (hence subdirectly irreducible, see also Jónsson [**53**]). (P3) follows from the analogous result for finite lattices and finite partition lattices, due to Pudlák and Tuma [**80**].

THEOREM 2.5 (McKenzie [**72**]). *Let $\mathcal{V}$ be a variety of algebras and consider the following statements about a subvariety $\mathcal{V}'$ of $\mathcal{V}$:*

(i) *$\mathcal{V}'$ is completely join prime in $\Lambda_{\mathcal{V}}$ (i.e. $\mathcal{V}' \leq \sum_{i \in I} \mathcal{V}_i$ implies $\mathcal{V}' \leq \mathcal{V}_i$ some $i \in I$);*

(ii) *$\mathcal{V}'$ can be generated by a finite subdirectly irreducible member;*

(iii) *$\mathcal{V}'$ is completely join irreducible in $\Lambda_{\mathcal{V}}$;*

(iv) *$\mathcal{V}'$ can be generated by a finitely generated subdirectly irreducible member;*

(v) *$\mathcal{V}'$ can be generated by a (single) subdirectly irreducible member;*

(vi) *$\mathcal{V}'$ is join irreducible in $\Lambda_{\mathcal{V}}$;*

Then we always have (iii)⇒(iv)⇒(v). If (P1) holds, then (i)⇒(ii), and if $\mathcal{V}$ is congruence distributive then (ii)⇒(iii) and (v)⇒(vi).

PROOF. (iii)⇒(iv) Every variety is generated by its finitely generated subdirectly irreducible members, so if $\mathcal{V}'$ is completely join irreducible, then it must be generated by one of them. (iv)⇒(v) is obvious.

Suppose now that $\mathcal{V} = (\mathcal{V}_F)^{\mathcal{V}}$ (i.e. (P1) holds). Then $\mathcal{V}$ is the join of all its finitely generated subvarieties. If $\mathcal{V}' \subseteq \mathcal{V}$ is completely join prime, then it is included in one of these, and therefore $\mathcal{V}'$ is itself finitely generated. This means that $\mathcal{V}'$ can be generated by finitely many finite subdirectly irreducible algebras, and since it is also join irreducible, it must be generated by just one of them, i.e. (ii) holds.

If $\mathcal{V}$ is congruence distributive and (ii) holds, then Theorem 2.3(i) implies that $\mathcal{V}'$ is join irreducible, and by Corollary 1.7(iii), $\mathcal{V}'$ has only finitely many subvarieties, hence it is completely join irreducible. (v)⇒(vi) also follows from Theorem 2.3(i). □

Thus for $\mathcal{V} = \mathcal{L}$ we have (i)⇒(ii)⇒(iii)⇒(iv)⇒(v)⇒(vi). McKenzie also gives examples of lattice varieties which show that, in general, none of the reverse implications hold. If $\mathcal{V}'$ is assumed to be finitely generated, then of course (ii)–(vi) are equivalent.

THEOREM 2.6 (Jónsson [**67**]). *Let V be a congruence distributive variety of algebras. Then*

(i) *(P1) implies that every proper subvariety of V has a cover in Λ_V;*

(ii) *(P2) implies that V is join irreducible in Λ_V;*

(iii) *(P1) and (P2) imply that V has no dual cover.*

PROOF. (i) If V' is a proper subvariety of $V = (V_F)^V$, then there exists a finite algebra $A \in V$ such that $A \notin V'$. By Theorem 2.3(iii) the quotient $\{A\}^V + V'/V'$ is finite and therefore contains a variety that covers V'.

(ii) If V' and V'' are proper subvarieties of V, then there exist algebras A' and A'' in V such that $A' \notin V'$ and $A'' \notin V''$. Assuming that $V \subseteq SV_{SI}$, we can find a subdirectly irreducible algebra $A \in V$ which has $A' \times A''$ as subalgebra. Then $A \notin V'$ and $A \notin V''$, so Theorem 2.3(i) implies that $A \notin V' + V''$, whence $V' + V'' \neq V$.

(iii) Again we let V' be a proper subvariety of V. By (P1) there exists a finite algebra $A \in V$ such that $A \notin V'$. Now (P2) implies that $\{A\}^V \neq V$, whence by (ii) V properly contains $V' + \{A\}^V$ which in turn properly contains V'. Consequently V' is not a dual cover of V.   □

The cardinality of Λ. Let $\mathcal{J}$ be the (countable) set of all lattice identities. Since every variety in Λ is defined by some subset of $\mathcal{J}$, we must have $|\Lambda| \leq 2^\omega$. The same argument shows that if V is any variety of algebras (of finite or countable type), then $|\Lambda_V| \leq 2^\omega$. Whether this upper bound on the cardinality is actually attained depends on the variety V. For $V = \mathcal{L}$, the answer is affirmative, as was proved independently by McKenzie [**70**] and Baker [**69**] (see also Wille [**72**] and Lee [**85**]). Baker in fact shows that the lattice $\Lambda_{\mathcal{M}}$ of all modular subvarieties contains the Boolean algebra 2^ω as a sublattice. We postpone the proof of this result until we have covered some theory of projective spaces in the next chapter. In Section 4.3 we give another result, from Lee [**85**], which shows that there are 2^ω almost distributive lattice varieties (to be defined). In contrast, Jónsson's Lemma implies that any finitely generated congruence distributive variety V has only finitely many subvarieties and therefore Λ_V is finite.

An as yet unsolved problem about lattice varieties is whether the converse of the above observation is true, i.e. if a lattice variety has only finitely many subvarieties, is it finitely generated? This problem can also be approached from below: if a lattice variety V is finitely generated, is every cover of V finitely generated?

Sometimes these problems are phrased in terms of the height of a variety V in Λ (= length of the ideal $(V]$). Since Λ is distributive, to be of finite height is of course equivalent to having only finitely many subvarieties. Call a variety V *locally finite* if every finitely generated member of V is finite. For locally finite congruence distributive varieties, the above problem is easily solved.

THEOREM 2.7 *Every finitely generated variety of algebras is locally finite. Conversely, if V is a locally finite congruence distributive variety, then*

(i) *every join irreducible subvariety of V that has finite height in Λ_V is generated by a finite subdirectly irreducible member, and*

(ii) *every variety of finite height in Λ_V is finitely generated.*

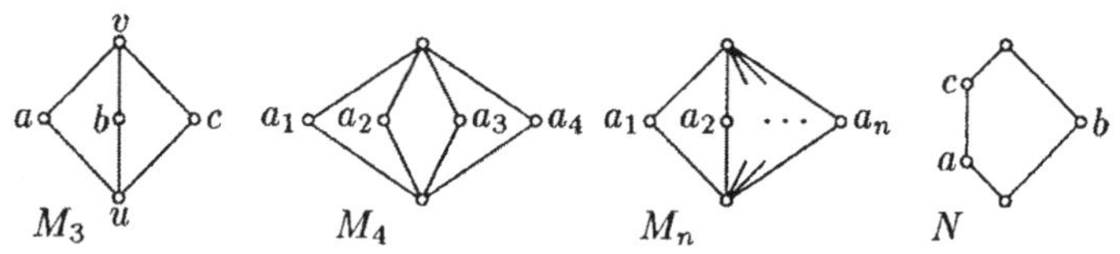

Figure 2.1

PROOF. If $\mathcal{V}$ is finitely generated, then $\mathcal{V} = \{A\}^{\mathcal{V}}$ for some finite algebra $A \in \mathcal{V}$. For $n \in \omega$ the elements of $F_{\mathcal{V}}(n)$ are represented by n-ary polynomial functions from A^n to A, of which we can have at most $|A|^{|A|^n}$. Hence $F_{\mathcal{V}}(n)$ is finite for each $n \in \omega$, and this is equivalent to $\mathcal{V}$ being locally finite.

Conversely, assume that $\mathcal{V}$ is a locally finite congruence distributive variety. (i) If a subvariety $\mathcal{V}'$ of $\mathcal{V}$ is join irreducible and has finite height in $\Lambda_{\mathcal{V}}$, then it is in fact completely join irreducible, whence Theorem 2.5 implies that $\mathcal{V}'$ is generated by a finitely generated subdirectly irreducible algebra, which must be finite. (ii) follows from (i) and the fact that a variety of finite height is the join of finitely many join irreducible varieties.   □

Nonfinitely based and nonfinitely generated varieties. It is easy to see that a variety can have at most countably many finitely based or finitely generated subvarieties, hence McKenzie's and Baker's result ($|\Lambda| = |\Lambda_{\mathcal{M}}| = 2^{\omega}$) imply that there are both nonfinitely based and nonfinitely generated (modular) lattice varieties. An example of a modular variety that is not finitely based is given in Section 5.3, and $\mathcal{L}$ and $\mathcal{M}$ are examples of varieties that are not finitely generated. In fact Freese [79] showed that, unlike $\mathcal{L}$, $\mathcal{M}$ is not even generated by its finite members (see Section 3.3). Other such varieties were previously discovered by Baker [69] and Wille [69].

2.2 The Structure of the Bottom of Λ

Covering relations between modular varieties. The class of all trivial (one-element) lattices, denoted by $\mathcal{T} = \mathrm{Mod}\{x = y\}$, is the smallest lattice variety and hence the least element of Λ. If a lattice variety $\mathcal{V}$ properly contains $\mathcal{T}$, then $\mathcal{V}$ must contain a lattice which has the two-element chain $\mathbf{2}$ as sublattice, hence $\mathbf{2} \in \mathcal{V}$. It is well known that, up to isomorphism, $\mathbf{2}$ is the only subdirectly irreducible distributive lattice, and therefore generates the variety of all distributive lattices, $\mathcal{D} = \{\mathbf{2}\}^{\mathcal{V}} = \mathrm{Mod}\{x(y+z) = xy + xz\}$. It follows that $\mathcal{D}$ is the unique cover of $\mathcal{T}$ in the lattice Λ.

The next important identity is the modular identity $xy + xz = x(y + xz)$, which defines the variety $\mathcal{M}$ of all modular lattices. Of course every distributive lattice is modular, and a lattice L satisfies the modular identity if and only if, for all $u, v, w \in L$ with $u \leq w$ we have $u + vw = (u + v)w$ (for arbitrary lattices we only have $\leq$ instead of equality). The *diamond* M_3 (see Figure 2.1) is the smallest example of a nondistributive modular lattice.

A well known result due to Birkhoff [35] states that M_3 is in fact a sublattice of every nondistributive modular lattice. We give a sketch of the proof. Let $x, y, z \in L$ and define $u = xy + xz + yz$ and $v = (x+y)(x+z)(y+z)$. Then clearly $u \leq v$, and the elements

$$a = u + xv = (u+x)v$$
$$b = u + yv = (u+y)v$$
$$c = u + zv = (u+z)v$$

generate a diamond with least element u and greatest element v. On the other hand, if $u = v$ for all choices of $x, y, z \in L$, then the identity

$$xy + xz + yz = (x+y)(x+z)(y+z)$$

holds in L, and it is not difficult to see that this identity is equivalent to the distributive identity.

It follows that every nondistributive modular lattice contains a sublattice isomorphic to M_3, and consequently the variety $\mathcal{M}_3 = \{M_3\}^{\mathcal{V}}$ covers $\mathcal{D}$. More generally, since the lattices M_n (see Figure 2.1) are simple (hence subdirectly irreducible) modular lattices for each $n \geq 3$, and since M_n is a sublattice of M_{n+1}, it follows from Corollary 1.7(i) that, up to isomorphism, $(\mathcal{M}_n)_{SI} = \{2\} \cup \{M_k : 3 \leq k \leq n\}$, where $\mathcal{M}_n = \{M_n\}^{\mathcal{V}}$. Hence the varieties $\mathcal{M}_n$ form a countable chain of join irreducible modular subvarieties of $\mathcal{M}$, with $\mathcal{M}_{n+1}$ covering $\mathcal{M}_n$ for $n \geq 3$. Jónsson [68] proved that for $n \geq 4$, $\mathcal{M}_{n+1}$ is in fact the only join irreducible cover of $\mathcal{M}_n$, and that $\mathcal{M}_3$ has exactly two join irreducible covers, $\mathcal{M}_{3^2}$ and $\mathcal{M}_4$. This result is presented in Section 3.4. Further remarks about the covers of $\mathcal{M}_{3^2}$ and various other modular varieties appear at the end of Chapter 3.

Covering relations between nonmodular varieties. A lattice variety is said to be *nonmodular* if it contains at least one nonmodular lattice L (i.e. $L \notin \mathcal{M}$). If $L \in \mathcal{V}$ is nonmodular, then we can infer the existence of three elements $u, v, w \in L$ such that $u \leq w$ and $u + vw < (u+v)w$. In that case the elements $a = u + vw$, $b = v$ and $c = (u+v)w$ generate a sublattice of L that is isomorphic to the *pentagon* N with critical quotient c/a (see Figure 2.1). Since the pentagon is nonmodular, one obtains the well known result of Dedekind [00]:

> *Every nonmodular lattice contains a sublattice isomorphic to N.*

Many of the later results will be of a similar flavor, in the sense that a certain property is shown to fail precisely because of the presence of some particular sublattices. If L and K are lattices, we say that L *excludes* K if L does not have a sublattice isomorphic to K. Otherwise we say that L *includes* K. In this terminology, modularity is said to be characterized by the exclusion of the pentagon.

An immediate consequence is that the variety generated by the pentagon (denoted by $\mathcal{N}$) is the smallest nonmodular variety. Again, Jónsson's Lemma enables us to compute $\mathcal{N}_{SI} = \{2, N\}$ and hence $\mathcal{N}$ is a join irreducible cover of the distributive variety $\mathcal{D}$. Since every lattice is either modular or nonmodular, we conclude that $\mathcal{M}_3$ and $\mathcal{N}$ are the only covers of $\mathcal{D}$.

In a paper of McKenzie [72] there is a list of 15 subdirectly irreducible lattices $L_1, L_2, \ldots, L_{15}$ (see Figure 2.2) with the following property: If we let $\mathcal{L}_i = \{L_i\}^{\mathcal{V}}$ ($i = 1, \ldots, 15$) then each of them satisfy $(\mathcal{L}_i)_{SI} = \{2, N, L_i\}$. Hence each $\mathcal{L}_i$ is a join irreducible cover of the variety $\mathcal{N}$. It is a nontrivial result, due to Jónsson and Rival [79], that McKenzie's list is complete. A proof of this result appears in Chapter 4.

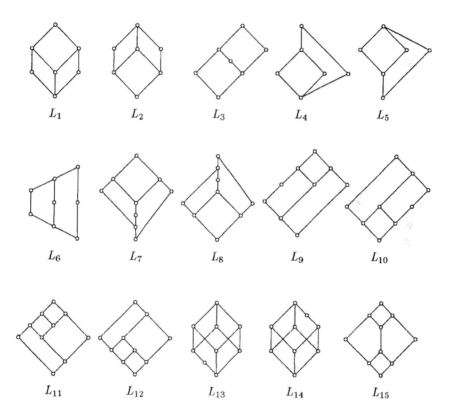

Figure 2.2

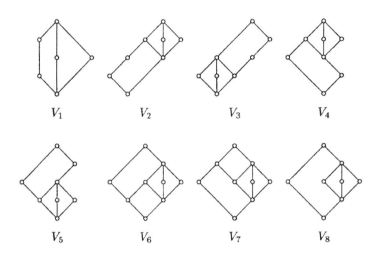

V_1 V_2 V_3 V_4

V_5 V_6 V_7 V_8

Figure 2.3

Rose [84] proves that above each of the varieties $\mathcal{L}_6$, $\mathcal{L}_7$, $\mathcal{L}_8$, $\mathcal{L}_9$, $\mathcal{L}_{10}$, $\mathcal{L}_{13}$, $\mathcal{L}_{14}$ and $\mathcal{L}_{15}$ there is a chain of varieties $\mathcal{L}_i^n$ ($n \in \omega$), each one generated by a finite subdirectly irreducible lattice L_i^n ($L_i^0 = L_i$), such that $\mathcal{L}_i^{n+1}$ is the unique join irreducible cover of $\mathcal{L}_i^n$ ($i = 6, 7, 8, 9, 10, 13, 14, 15$).

Lattice varieties which do not include any of the lattices $M_3, L_1, \ldots, L_{12}$ are called *almost distributive* by Lee [85]. They are all locally finite, and Lee shows that their finite subdirectly irreducible members can be characterized in a certain way which, in principle, enables us to determine the position of any finitely generated almost distributive variety in the lattice Λ.

Ruckelshausen [78] investigates the covers of $\mathcal{M}_3 + \mathcal{N}$, and further results by Nation [85] [86] include a complete list of the covers of the varieties $\mathcal{L}_1$ and $\mathcal{L}_{11}$, $\mathcal{L}_{12}$. Nation also shows that above $\mathcal{L}_{11}$ and $\mathcal{L}_{12}$ there are exactly two covering chains of join irreducible varieties. These results are discussed in more detail at the end of Chapter 4.

A diagram of Λ is shown in Figure 2.4.

2.3   Splitting Lattices and Bounded Homomorphisms

The concept of splitting. A pair of elements (x, y) in a lattice L is said to be a *splitting pair of* L if L is the disjoint union of the principal ideal $(x]$ and the principal filter $[y)$ (or equivalently, if for any $z \in L$ we have $z \leq x$ if and only if $y \nleq z$). The following lemma is an immediate consequence of this definition.

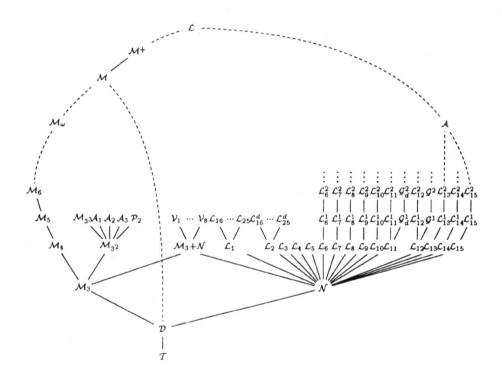

Figure 2.4

LEMMA 2.8 *In a complete lattice L the following conditions are equivalent:*

(i) (x, y) *is a splitting pair of L;*

(ii) *x is completely meet prime in L and $y = \prod_{z \not\leq x} z$;*

(iii) *y is completely join prime in L and $x = \sum_{z \not\geq y} z$.*

The notion of "splitting" a lattice into a (disjoint) ideal and filter was originally investigated by Whitman [**43**]. In McKenzie [**72**] this concept is applied to the lattice Λ, as a generalization of the familiar division of Λ into a modular and a nonmodular part. What is noteworthy about McKenzie's and subsequent investigations by others is that they yield greater insight, not only into the structure of Λ, but also into the structure of free lattices. In this section we first present some basic facts about splitting pairs of varieties in general and then discuss some related concepts and their implications for lattices.

Let $\mathcal{V}$ be a variety of algebras and suppose $(\mathcal{V}_0, \mathcal{V}_1)$ is a splitting pair in Λ. By Lemma 2.8 $\mathcal{V}_0$ is completely meet prime, hence completely meet irreducible, and since

$$\mathcal{V}_0 = \text{Mod Id} \, \mathcal{V}_0 = \bigcap_{\varepsilon \in \text{Id}\mathcal{V}_0} \text{Mod}\{\varepsilon\}$$

it follows that $\mathcal{V}_0$ can be defined by a single identity ε_0. Dually, since every variety is generated by its finitely generated subdirectly irreducible members, $\mathcal{V}_1 = \{A\}^{\mathcal{V}}$ for some finitely generated subdirectly irreducible algebra A. We shall refer to such an algebra A (which generates a completely join prime subvariety of $\mathcal{V}$) as a *splitting algebra* in $\mathcal{V}$, and to the variety $\mathcal{V}_0$ as its *conjugate variety*, defined by the *conjugate identity* ε_0. Note that if $\mathcal{V}$ is generated by its finite members, then we can assume, by Theorem 2.5, that A is a finite algebra. If, in addition, $\mathcal{V}$ is congruence distributive, then Corollary 1.7(iv) implies that A is unique. In particular, if $(\mathcal{V}_0, \mathcal{V}_1)$ is a splitting pair in Λ, then $\mathcal{V}_1 = \{L\}^{\mathcal{V}}$ where L is a finite subdirectly irreducible lattice, and we refer to such a lattice as a *splitting lattice*. The two standard examples of splitting pairs in Λ are $(\mathcal{T}, \mathcal{D})$ and $(\mathcal{M}, \mathcal{N})$.

Projective Algebras. An algebra P in a class $\mathcal{K}$ of algebras said to be *projective in $\mathcal{K}$* if for any homomorphism $h : P \to B$ and epimorphism $g : A \twoheadrightarrow B$ with $A, B \in \mathcal{K}$, there exists a homomorphism $f : P \to A$ such that $h = gf$ (Figure 2.5(i)).

An algebra B is a *retract* of an algebra A if there exist homomorphisms $f : B \to A$ such that gf is the identity on B. Clearly f must be an embedding, and g is called a *retraction* of f.

LEMMA 2.9 *Let $\mathcal{K}$ be a class of algebras in which $\mathcal{K}$-free algebras exist. Then, for any $P \in \mathcal{K}$, the following conditions are equivalent:*

(i) *P is projective in $\mathcal{K}$;*

(ii) *For any algebra $A \in \mathcal{K}$ and any epimorphism $g : A \twoheadrightarrow P$ there exists an embedding $f : P \hookrightarrow A$ such that gf is the identity on P;*

(iii) *P is a retract of a $\mathcal{K}$-free algebra.*

PROOF. (ii) is a special case of (i), with $B = P$ and h the identity on P. Clearly f must be an embedding in this case. Suppose (ii) holds, and let X be a set with $|X| = |P|$. Then

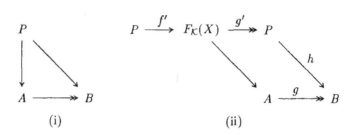

Figure 2.5

there exists an epimorphism $g : F_{\mathcal{K}}(X) \twoheadrightarrow P$. Since $F_{\mathcal{K}}(X) \in \mathcal{K}$, it follows from (ii) that P is a retract of $F_{\mathcal{K}}(X)$. Lastly, suppose P is a retract of some K-free algebra $F_{\mathcal{K}}(X)$, and let $h : P \to B$ and $g : A \twoheadrightarrow B$ be given $(A, B \in \mathcal{K})$. Then there exist $f' : P \to F_{\mathcal{K}}(X)$ and $g' : F_{\mathcal{K}}(X) \to P$ such that $g'f'$ is the identity on P (Figure 2.5(ii)). Since g is onto, we can define a map $k : X \to A$ satisfying $gk(x) = hg'(x)$ for all $x \in X$. Let $\overline{k}$ be the extension of k to $F_{\mathcal{K}}(X)$, then $\overline{k}f'$ is the required homomorphism from P to A.   □

Thus, for any variety $\mathcal{V}$, every $\mathcal{V}$-free algebra is projective in $\mathcal{V}$.

LEMMA 2.10 (Rose [84]). *Let $\mathcal{K}$ be a class of algebras and suppose that $P \in \mathcal{K}^{\mathcal{V}}$ is subdirectly irreducible and projective in $\mathcal{K}^{\mathcal{V}}$. Then P is isomorphic to a subalgebra of some member of $\mathcal{K}$.*

PROOF. Since $P \in \mathcal{K}^{\mathcal{V}} = \mathbf{HSP}\mathcal{K}$, Lemma 2.9 (ii) implies that $P \in \mathbf{SP}\mathcal{K}$. Hence we can assume that P is a subalgebra of a direct product $\chi_{i \in I} A_i$, where $A_i \in \mathcal{K}$ for i in some index set I. Denoting the projection map from the product to each A_i by π_i, we see that P is a subdirect product of the family of algebras $\{\pi_i(P) : i \in I\}$. But P is assumed to be subdirectly irreducible, so there exists $j \in I$ such that $\pi_i(P)$ is isomorphic to P. Therefore $\pi_j : P \to L_j$ is an embedding.   □

Recall from Section 1.1 that $F_{\mathcal{V}}(X)$ can be constructed as a quotient algebra of the word algebra $W(X)$, whence every element of $F_{\mathcal{V}}(X)$ can be represented by a term of $W(X)$. Also if A is an algebra and p, q are terms in $W(X)$, then the identity $p = q$ holds in A if and only if $h(p) = h(q)$ for every homomorphism $h : W(X) \to A$. Notice that if $\mathcal{V}$ is a variety containing A, then any such h can be factored through $F_{\mathcal{V}}(X)$. The following theorem was proved by McKenzie [72] for $\mathcal{L}$-free lattices, and then generalized to projective lattices by Wille [72] and to projective algebras by Day [75].

THEOREM 2.11 *Let $\mathcal{V}$ be a variety of algebras, suppose $P \in \mathcal{V}$ is projective in $\mathcal{V}$, and for some $a, b \in P$ there is a largest congruence $\psi \in \mathrm{Con}(P)$ which does not identify a and b. Then P/ψ is a splitting algebra in $\mathcal{V}$, and if $f : P \hookrightarrow F_{\mathcal{V}}(X)$ is an embedding and $g : F_{\mathcal{V}}(X) \twoheadrightarrow P$ is a retraction of f (i.e. $gf = \mathrm{id}_P$) then for any terms p, q which represent $f(a), f(b)$ respectively, the identity $p = q$ is a conjugate identity of P/ψ.*

PROOF. It is enough to show that $p = q$ fails in P/ψ but holds in any subvariety of $\mathcal{V}$ that does not contain P/ψ. Let $\gamma : P \to P/\psi$ be the canonical epimorphism. Then γg does not identify $f(a)$ and $f(b)$, hence by the remark above, $p = q$ fails in P/ψ. Suppose now that $\mathcal{V}'$ is any subvariety of $\mathcal{V}$ not containing P/ψ, and let $h : F_{\mathcal{V}}(X) \twoheadrightarrow F_{\mathcal{V}'}(X)$ be the extension of the identity map on the generating set X. Clearly the identity $p = q$ will hold in $\mathcal{V}'$ if and only if $hf(a) = hf(b)$. Suppose to the contrary that a and b are not identified by $\ker hf$. Since ψ is assumed to be the largest such congruence, we have $\ker hf \subseteq \psi$, and it follows from the second homomorphism theorem that $\gamma : P \twoheadrightarrow P/\psi$ factors through $F_{\mathcal{V}'}(X)$, hence $P/\psi \in \mathcal{V}'$, a contradiction. Therefore $p = q$ holds in $\mathcal{V}'$.   □

In particular, any projective subdirectly irreducible algebra is splitting. Combining Lemma 1.10 with the above theorem we obtain the result of Wille [**72**]:

COROLLARY 2.12 *Let $\mathcal{V}$ be a lattice variety, u/v a prime quotient in some lattice $P \in \mathcal{V}$ which is projective in $\mathcal{V}$, and suppose θ is the largest congruence on P that does not collapse u/v. Then P/θ is a splitting lattice in $\mathcal{V}$.*

If we take $\mathcal{V}$ to be the variety $\mathcal{L}$ of all lattices, then the converse of the above corollary is also true. In fact it follows from a result of McKenzie (Corollary 2.26) that every splitting lattice in $\mathcal{L}$ is isomorphic to a quotient lattice of some finitely generated free lattice $F(n)$ modulo a congruence θ which is maximal with respect to not collapsing some prime quotient of $F(n)$. Before we can prove this result, however, we need some more information about free lattices, which is due to Whitman [**41**] [**42**] and can also be found in [**ATL**] and [**GLT**]. Whitman showed that a lattice L is ($\mathcal{L}$-) freely generated by a set $X \subseteq L$ if and only if for all $x, y \in X$ and $a, b, c, d \in L$ the following four conditions are satisfied:

(W1) $x \leq y$ implies $x = y$ (i.e. generators are incomparable)
(W2) $ab \leq y$ implies $a \leq y$ or $b \leq y$
(W3) $x \leq c + d$ implies $x \leq c$ or $x \leq d$
(W) $ab \leq c + d$ implies $a \leq c + d$ or $b \leq c + d$ or $ab \leq c$ or $ab \leq d$.

(These conditions are also known as Whitman's solution to the word problem for lattices since they provide an algorithm for testing when two lattice terms represent the same element of a free lattice.)

In fact Jónsson [**70**] showed that if $\mathcal{V}$ is a nontrivial lattice variety then (W1), (W2) and (W3) hold in any $\mathcal{V}$-freely generated lattice. We give a proof of this result. A subset X of a lattice L is said to be *irredundant* if for any distinct $x, x_1, x_2, \ldots, x_n \in X$

(W2′) $x \not\geq x_1 x_2 \ldots x_n$ and
(W3′) $x \not\leq x_1 + x_2 + \ldots + x_n$.

We also require the important notion of a join-cover. Let U and V be two nonempty finite subsets of a lattice L. We say that U *refines* V (in symbols $U \ll V$) if for every $u \in U$ there exists $v \in V$ such that $u \leq v$. V is a *join-cover* of $a \in L$ if $a \leq \sum V$, and a join-cover V of a is *nontrivial* if $a \not\leq v$ for all $v \in V$. Observe that any join-cover of a, which refines a nontrivial join-cover of a, is itself nontrivial. The notion of a *meet-cover* is defined dually.

LEMMA 2.13 (Jónsson[**70**]). *Let L be a lattice generated by a set $X \subseteq L$.*

(i) *If every join-cover $U \subseteq X$ of an element $a \in L$ is trivial, then every join-cover of a is trivial.*

(ii) *X is irredundant if and only if for all $x, y \in X$ and $a, b, c, d \in L$ (W1), (W2) and (W3) hold.*

PROOF. (i) Let Y be a subset of L for which every join-cover $U \subseteq Y$ of a is trivial, and let $S(Y)$ and $P(Y)$ denote the sets of all elements of L that are (finite) joins and meets of elements of Y respectively. Since every join-cover $V \subseteq S(Y)$ of a is refined by a join-cover $U \subseteq Y$, V is trivial. We claim that the same holds for every join-cover $V \subseteq P(Y)$ of a. Suppose V is a finite nonempty subset of $P(Y)$ such that for all $v \in V$, $a \not\leq v$. Each element v is the meet of a nonempty set $U_v \subseteq Y$. Since $a \not\leq v$, there exists an element $u_v \in U_v$ such that $a \not\leq u_v$. Each u_v belongs to Y, so the set $W = \{u_v : v \in V\}$ cannot be a join-cover of a, i.e. $a \not\leq \sum W$. But $v = \prod U_v \leq u_v$, and hence $\sum V \leq \sum W$. It follows that $a \not\leq \sum V$ which means that V is not a join-cover of a. This contradiction proves the claim.

Now let $Y_0 = P(X)$ and $Y_{n+1} = PS(Y_n)$ for $n \in \omega$. If every join-cover $U \subseteq X$ of a is trivial, then by the above this is also true for X replaced by Y_n. Since X generates L, we have $L = \bigcup_{n \in \omega} Y_n$, so if V is any join-cover of a, then $V \subseteq Y_m$ for some $m \in \omega$, and therefore V is trivial.

(ii) If X is irredundant, then clearly the elements of X must be incomparable, so (W1) is satisfied. It also follows that any join-cover $U \subseteq X$ of a generator $x \in X$ is trivial, hence by part (i) every join-cover of x is trivial. This implies (W3), and (W2) follows by duality. Conversely, if $x \leq x_1 + x_2 + \cdots + x_n$ with all $x, x_1, \ldots, x_n \in X$ distinct, then repeated application of (W3) yields $x \leq x_i$ and by (W1) $x = x_i$ for some $i = 1, \ldots, n$. This contradiction, and its dual argument, shows that X is irredundant. □

THEOREM 2.14 (Jónsson[**70**]). *Let $\mathcal{K}$ be a class of lattices that contains at least one nontrivial lattice. If F is a lattice that is $\mathcal{K}$-freely generated by a set X, then X is irredundant.*

PROOF. Suppose $x, x_1, \ldots, x_n \in X$ are distinct, and let $L \in \mathcal{K}$ be a lattice with more than one element. Then there exists $a, b \in L$ such that $a \not\leq b$. Choose a map $f : X \to L$ such that $f(x) = a$ and $f(x_i) = b$ (all i). Then the extension of f to a homomorphism $\overline{f} : F \to L$ satisfies $\overline{f}(x) = a \not\leq b = \overline{f}(x_1 + \ldots + x_n)$, and hence $x \not\leq x_1 + \ldots + x_n$. By duality X is irredundant. □

The last condition (W) is usually referred to as Whitman's condition, and it may be considered as a condition applicable to lattices in general, since it makes no reference to the generators of L. Clearly, if a lattice L satisfies (W), then every sublattice of L again satisfies (W). In particular Whitman's result and Lemma 2.9 show that every projective lattice satisfies (W). Day [**70**] found a very simple proof of this fact based on a construction which we will use several times in this section and in Chapters 4 and 6. Given a lattice L and a quotient $I = u/v$ in L, we construct a new lattice

$$L[I] = (L - I) \cup (I \times \mathbf{2})$$

with the ordering $x \leq y$ in L if and only if one of the following conditions holds:

(i) $x, y \in L - I$ and $x \leq y$ in L;

(ii) $x \in L - I$, $y = (b, j)$ and $x \leq b$ in L;

(iii) $y \in L - I$, $x = (a, i)$ and $y \leq a$ in L;

(iv) $x = (a, i)$, $y = (b, j)$ and $a \leq b$ in L, $i \leq j$ in $\mathbf{2}$.

$L[I]$ is referred to as L *with doubled quotient* u/v, and it is easy to check that $L[I]$ is in fact a lattice ($\mathbf{2} = \{0, 1\}$ is the two-element chain with $0 < 1$). Also there is a natural epimorphism $\gamma : L[u/v] \twoheadrightarrow L$ defined by

$$\gamma(x) = \begin{cases} x & \text{if} \quad x \in L - u/v \\ a & \text{if} \quad x = (a, i) \quad \text{some} \quad i \in \mathbf{2}. \end{cases}$$

We say that a 4-tuple $(a, b, c, d) \in L^4$ a (W)-*failure* of L if $ab \leq c + d$ but $ab \not\leq c, d$ and $a, b \not\leq c + d$.

LEMMA 2.15 (Day [70]). *Let (a, b, c, d) be a (W)-failure of L, and let $u/v = c + d/ab$. Then there does not exist an embedding $f : L \hookrightarrow L[u/v]$ such that γf is the identity map on L (i.e. there exists no coretraction of γ).*

PROOF. Suppose the contrary. Then $f(x) = x$ for each $x \in L - u/v$, and $f(v) \leq f(u)$. But $f(v) = f(ab) = f(a)f(b) = ab = (v, 1)$ since $a, b \not\leq u$, and dually $f(u) = (u, 0)$. This is a contradiction since by definition $(v, 1) \not\leq (u, 0)$.   □

From the equivalence of (i) and (ii) of Lemma 2.9 we now obtain:

COROLLARY 2.16

(i) *Every lattice which is projective in $\mathcal{L}$ satisfies (W).*

(ii) *Every free lattice satisfies (W1), (W2), (W3) and (W).*

As mentioned before, the converse of (ii) is also true. A partial converse of (i) is given by Theorem 2.19.

Bounded homomorphisms. A lattice homomorphism $f : L \to L'$ is said to be

upper bounded — if for every $b \in L'$ the set $f^{-1}(b] = \{x \in L : f(x) \leq b\}$ is either empty
    or has a greatest element, denoted by $\alpha_f(b)$;

lower bounded — if for every $b \in L'$ the set $f^{-1}[b) = \{x \in L : f(x) \geq b\}$ is either empty
    or has a least element, denoted by $\beta_f(b)$;

bounded — if f is both upper and lower bounded;

(upper/lower) bounded — if a given one of the above three properties holds.

The following lemma lists some easy consequences of the above definitions.

LEMMA 2.17 *Let $f : L \to L'$ and $g : L' \to L''$ be two lattice homomorphisms.*

(i) *If L is a finite lattice, then f is bounded.*

(ii) *If f and g are (upper/lower) bounded then gf is (upper/lower) bounded.*

(iii) If gf is (upper/lower) bounded and f is an epimorphism, then g is (upper/lower) bounded.

(iv) If gf is (upper/lower) bounded and g is an embedding, then f is (upper/lower) bounded.

(v) If f is an upper bounded epimorphism and $b \in L'$ then $\alpha_f(b)$ is the greatest member of $f^{-1}\{b\}$, and the map $\alpha_f : L' \hookrightarrow L$ is meet preserving. Dually, if f is a lower bounded epimorphism then $\beta_f(b)$ is the least member of $f^{-1}\{b\}$, and $\beta_f : L' \hookrightarrow L$ is join preserving.

A lattice is said to be *upper bounded* if it is an upper bounded epimorphic image of some free lattice, and *lower bounded* if the dual condition holds. A lattice which is a bounded epimorphic image of some free lattice is said to be *bounded* (not to be confused with lattices that have a largest and a smallest element: such lattices will be referred to as 0,1 - lattices). Of course every bounded lattice is both upper and lower bounded. We shall see later (Theorem 2.23) that, for finitely generated lattices, the converse also holds.

The notion of a bounded homomorphism was introduced by McKenzie [**72**], and he used it to characterize splitting lattices as subdirectly irreducible finite bounded lattices (Theorem 2.25). We first prove a result of Kostinsky [**72**] which shows that every bounded lattice that satisfies Whitman's condition (W) is projective. For finitely generated lattices this result was already proved in McKenzie [**72**].

LEMMA 2.18 *Suppose $F(X)$ is a free lattice generated by the set X and let $f : F(X) \to L$ be a lower bounded epimorphism. Then for each $b \in L$ the set $\{x \in X : f(x) \geq b\}$ is finite.*

PROOF. Since f is lower bounded and onto, $\beta_f(b)$ exists. Suppose it is represented by a term $t(x_1, \ldots, x_n) \in F(X)$ for some $x_1, \ldots, x_n \in X$, then clearly

$$\{x \in X : f(x) \geq b\} = \{x \in X : x \geq t(x_1, \ldots, x_n)\} \subseteq \{x_1, \ldots, x_n\}.$$

□

In particular note that if a finite lattice is a (lower) bounded homomorphic image of a free lattice F, then F must necessarily be finitely generated.

THEOREM 2.19 (Kostinsky [**72**]). *Every bounded lattice that satisfies (W) is projective (in $\mathcal{L}$).*

PROOF. Let f be a bounded homomorphism from some free lattice $F(X)$ onto a lattice L, and suppose that L satisfies (W). We show that L is a retract of $F(X)$, and then apply Lemma 2.9. To simplify the notation we will denote the maps α_f and β_f simply by α and β.

Let $h : F(X) \to F(X)$ be the endomorphism that extends the map $x \mapsto \alpha f(x)$, $x \in X$. We claim that for each $a \in F(X)$, $\beta f(a) \leq h(a) \leq \alpha f(a)$, from which it follows that $fh(a) = f(a)$. This is clearly true for $x \in X$. Suppose it holds for $a, a' \in F(X)$. Since β is join-preserving

$$\begin{aligned}\beta f(a + a') = \beta f(a) + \beta f(a') &\leq h(a) + h(a') \\ &= h(a + a') \leq \alpha f(a) + \alpha f(a') \leq \alpha f(a + a')\end{aligned}$$

and similarly since α is meet-preserving $\beta f(aa') \leq h(aa') \leq \alpha f(aa')$. This establishes the claim.

We now define the map $g : L \to F(X)$ by $g = h\beta$. Then g is clearly join-preserving. Also $fg(b) = fh\beta(b) = f\beta(b) = b$ for all $b \in L$ (since $fh = f$). Thus we need only show that g is meet-preserving. Since $h\beta$ is orderpreserving, it suffices to show that

$$(*) \qquad h\beta(bc) \geq h(\beta(b)\beta(c)) \qquad \text{for all} \quad b, c \in L.$$

We first observe that, by the preceding lemma, the set $S = \{x \in X : f(x) \geq bc\}$ is finite. Let

$$u = \begin{cases} \beta(b)\beta(c) \prod S & \text{if} \quad S \neq \emptyset \\ \beta(b)\beta(c) & \text{if} \quad S = \emptyset. \end{cases}$$

Then clearly $f(u) = bc$. We claim that $u = \beta(bc)$. To see this, let Y be the set of all $a \in F(X)$ such that $bc \leq f(a)$ implies $u \leq a$. By definition $X \subseteq Y$, and Y is closed under meets. Suppose $a, a' \in Y$ and $bc \leq f(a) + f(a')$. Since L satisfies (W), we have

$$b \leq f(a) + f(a') \quad \text{or} \quad c \leq f(a) + f(a') \quad \text{or} \quad bc \leq f(a) \quad \text{or} \quad bc \leq f(a').$$

Applying β to the first two cases we obtain $u \leq \beta(b) \leq a + a'$ or $u \leq \beta(c) \leq a + a'$, and in the third and fourth case we have $u \leq a \leq a + a'$ or $u \leq a' \leq a + a'$ (since $a, a' \in Y$). Thus $Y = F(X)$, and it follows that u is the least element for which $f(u) \geq bc$, i.e. $u = \beta(bc)$.

Now consider $h(u) = h\beta(bc)$. If $S = \emptyset$ then $\beta(bc) = \beta(b)\beta(c)$ and so $(*)$ holds. If $S \neq \emptyset$ then $h\beta(bc) = h\beta(b)h\beta(c) \prod h(S)$ so to prove $(*)$, it is enough to show that $h(\beta(b)\beta(c)) \leq h(x)$ for all $x \in S$. But any such x satisfies $f(\beta(b)\beta(c)) = bc \leq f(x)$, and applying α we get $\beta(b)\beta(c) \leq \alpha f(x) = h(x)$. Thus $h(\beta(b)\beta(c)) \leq hh(x)$ and, since we showed that $h(a) \leq \alpha f(a)$, we have $hh(x) \leq \alpha f h(x) = \alpha f(x) = h(x)$ as required. $\square$

A complete characterization of the projective lattices in $\mathcal{L}$ can be found in Freese and Nation [**78**].

We now describe a particularity elegant algorithm, due to Jónsson, to determine whether a finitely generated lattice is bounded.

Let $D_0(L)$ be the set of all $a \in L$ that have no nontrivial join-cover (i.e. the set of all join-prime elements of L). For $k \in \omega$ let $D_{k+1}(L)$ be the set of all $a \in L$ such that if V is any nontrivial join-cover of a, then there exists a join-cover $U \subseteq D_k(L)$ of a with $U \ll V$. Note that $D_0(L) \subseteq D_1(L)$, and if we assume that $D_{k-1}(L) \subseteq D_k(L)$ for some $k \geq 1$, then for any $a \in D_k(L)$ any nontrivial join-cover V of a there exists a join-cover $U \subseteq D_{k-1}(L) \subseteq D_k(L)$ of a with $U \ll V$, whence $a \in D_{k+1}(L)$. So, by induction we have

$$D_0(L) \subseteq D_1(L) \subseteq \ldots \subseteq D_k(L) \subseteq \ldots.$$

Finally, let $D(L) = \bigcup_{k\in\omega} D_k(L)$ and define the sets $D'_k(L)$ and $D'(L)$ dually. Jónsson's algorithm states that a finitely generated lattice is bounded if and only if $D(L) = L = D'(L)$. This result will follow from Theorem 2.23.

LEMMA 2.20 (Jónsson and Nation [**75**]). *Suppose L is a lattice generated by a set $X \subseteq L$, let H_0 be the set of all (finite) meets of elements of X, and for $k \in \omega$ let H_{k+1} be the set of all meets of joins of elements of H_k. Then $D_k(L) \subseteq H_k$ for all $k \in \omega$.*

PROOF. Note that the H_k are closed under meets, $H_0 \subseteq H_1 \subseteq \ldots$ and $L = \bigcup_{k\in\omega} H_k$. Suppose $a \in D_0(L)$ and let $m \in \omega$ be the smallest number for which $a \in H_m$. If $m > 0$, then a is of the form $a = \prod_{i=1}^{n} \sum U_i$, where each U_i is a finite subset of H_{m-1}. U_i is a join-cover of a for each $i = 1, \ldots, n$, and since $a \in D_0(L)$, it must be trivial. Hence there exists $a_i \in U_i$ with $a \le a_i$ for each i. But then $a = \prod_{i=1}^{n} a_i \in H_{m-1}$, a contradiction. Thus $D_0(L) \subseteq H_0$.

We proceed by induction. Suppose $D_k(L) \subseteq H_k$, $a \in D_{k+1}(L)$ and $m \ge k+1$ is the smallest number for which $a \in H_m$. Again a is of the form $a = \prod_{i=1}^{n} \sum U_i$ with $U_i \subseteq H_{m-1}$ and each U_i is a join-cover of a. If U_i is trivial, pick $a_i \in U_i$ with $a \le a_i$, and if U_i is nontrivial, pick a join-cover $V_i \subseteq D_k(L)$ of a with $V_i \ll U_i$. By assumption each V_i is a subset of H_k. If $m > k+1$ then $\sum V_i \in H_{k+1}$ and a is the meet of these elements $\sum V_i$ and the a_i, so $a \in H_{m-1}$, a contradiction. Thus $m = k+1$ and $D_{k+1}(L) \subseteq H_{k+1}$. $\quad\square$

For the next lemma, note that Whitman's condition (W) is equivalent to the following: for any two finite subset U, V of L, if $a = \prod U \le \sum V = b$, then V is a trivial join-cover of a or U is a trivial meet-cover of b.

LEMMA 2.21 *Suppose $L = F(X)$ is freely generated by X, and let H_k be as in the previous lemma. Then $D_k(L) = H_k$ and therefore $D(L) = L$.*

PROOF. By the previous lemma, it is enough to show that $H_k \subseteq D_k(L)$ for each $k \in \omega$. If $a \in H_0$, then $a = \prod_{i=1}^{n} x_i$ for some $x_i \in X$, and Whitman's condition (W) and (W1) imply that any join-cover of a must be trivial, hence $a \in D_0(L)$. Next suppose $a \in H_1$. Then $a = \prod_{i=1}^{n} \sum U_i$ for some finite sets $U_i \subseteq H_0 = D_0(L)$, some $n \in \omega$. If V is a nontrivial join-cover of a, then (W) implies that for some i_0 we have $\sum U_{i_0} \le \sum V$ (see remark above). Since $U_{i_0} \subseteq D_0(L)$, V is a trivial join-cover of each $u \in U_{i_0}$, and therefore $U_{i_0} \ll V$. Since U_{i_0} is also a join-cover of a, it follows that $a \in D_1(L)$.

Proceeding by induction, suppose now that $D_k(L) = H_k$ and $a \in H_{k+1}$, for some $k \ge 1$. Then $a = \prod_{i=1}^{n} \sum U_i$ for some $U_i \subseteq H_k$, some $n \in \omega$, and each U_i is a join-cover of a. Let V be any nontrivial join-cover of a. As before (W) implies that $\sum U_{i_0} \le \sum V$ for some i_0. Let W be the set of all $u \in U_{i_0}$ such that V is a nontrivial join-cover of u, and set $W' = U_{i_0} - W$. Since $W \subseteq H_k = D_k(L)$, there exists for each $u \in W$ a join-cover $V_u \subseteq D_{k_1}(L)$ of u with $V_u \ll V$. It is now easy to check that $U = W' \cup \bigcup_{u \in W} V_u$ is a join-cover of a which refines V and is contained in $D_k(L)$. Hence $a \in D_{k+1}(L)$. $\quad\square$

A join-cover V of a in a lattice L is said to be *irredundant* if no proper subset of V is a join-cover of a, and *minimal* if for any join-cover U of a, $U \ll V$ implies $V \subseteq U$. Observe that every join-cover contains an irredundant join-subcover (since it is a finite set) and that the elements of an irredundant join-cover are noncomparable. Also, every minimal join-cover is irredundant.

LEMMA 2.22 (Jónsson and Nation [75]). *If F is a free lattice and V is a nontrivial cover of some $a \in D_k(F)$, then there exists a minimal cover V_0 of a with $V_0 \ll V$ and $V_0 \subseteq D_{k-1}(F)$.*

PROOF. First assume that F is freely generated by a finite set X. Suppose V is a nontrivial join-cover of $a \in D_k(F)$, and let $\mathcal{C}$ be the collection of all irredundant join-covers $U \subseteq D_{k-1}(F)$ of a, which refine V. $\mathcal{C}$ is nonempty since $a \in D_k(F)$, and by Lemma 2.20 $D_k(F)$ is finite, hence $\mathcal{C}$ is finite. Note that if $U \in \mathcal{C}$ and $U \ll W \ll U$ for some subset W of F, then for each $u \in U$ there exists $w \in W$ and $u' \in U$ such that $u \le w \le u'$, and since the elements of U are noncomparable, we must have $u = w = u'$ and therefore $U \subseteq W$. In particular, it follows that $\mathcal{C}$ is partially ordered by the relation $\ll$. Let V_0 be a minimal (with respect to $\ll$) member of $\mathcal{C}$, and suppose U is any join-cover of a with $U \ll V_0$. Because V is assumed to be nontrivial, so are V_0 and U, and since $a \in D_k(F)$, there exists a join-cover $U_0 \subseteq D_{k-1}(F)$ of a with $U_0 \ll U$. It follows that $U_0 \ll V_0$, and since we may assume that $U_0 \in \mathcal{C}$, we have $U_0 = V_0$. Therefore $V_0 \ll U \ll V_0$ which implies $V_0 \subseteq U$. Thus V_0 is a minimal join-cover of a.

Assume now that X is infinite, and choose a finite subset Y of X such that $V \cup \{a\}$ belong to the sublattice F' generated by $Y \subseteq F$. By the first part of the proof, there exists a set $V_0 \subseteq D_{k-1}(F') \subseteq D_{k-1}(F)$ such that $V_0 \ll V$ and V_0 is a minimal cover of a in F'. We show that V_0 is also a minimal cover of a in F. Let F_0 be the lattice obtained by adjoining a smallest element 0 to F, and let h be the endomorphism of F_0 that maps each member of Y onto itself and all the remaining elements of X onto 0. Then h maps every member of F' onto itself, and $h(u) \le u$ for all $u \in F_0$. Hence, if U is any join-cover of a in F', then the set $U' = h(U) - \{0\}$ is a join-cover of a in F' and $U' \ll V_0$, so that $V_0 \subseteq U' \ll U$. Thus $V_0 \ll U \ll V_0$, which implies that $V_0 \subseteq U$. $\qquad\square$

For finitely generated lattices we can now give an internal characterization of lower boundedness. This result, together with its dual, implies that an upper and lower bounded finitely generated lattice is bounded.

THEOREM 2.23 (Jónsson and Nation [75]). *For any finitely generated lattice L, the following statements are equivalent:*

(i) *L is lower bounded;*

(ii) *$D(L) = L$;*

(iii) *Every homomorphism of a finitely generated lattice into L is lower bounded.*

PROOF. Suppose (i) holds, let f be a lower bounded epimorphism that maps some free lattice F onto L, and denote by $\beta(a) = \beta_f(a)$ the smallest element of the set $f^{-1}[a]$ for all $a \in L$. We show by induction that

$$\beta(a) \in D_k(F) \qquad \text{implies} \qquad a \in D_k(L)$$

then $D(L) = L$ follows from the result $D(F) = F$ of Lemma 2.21. Since β is a join-preserving map, the image $\beta(U)$ of a join-cover U of a is a join-cover of $\beta(a)$. If $\beta(U)$ is trivial, then $\beta(a) \le \beta(u)$ for some $u \in U$, hence $a = f\beta(a) \le f\beta(u) = u$ and U is also trivial. It follows that $\beta(a) \in D_0(F)$ implies $a \in D_0(L)$.

Suppose now that $\beta(a) \in D_k(f)$ and that U is a nontrivial join-cover of a. Then $\beta(U)$ is a nontrivial join-cover of $\beta(a)$ and by Lemma 2.22 there exists a minimal join-cover $U_0 \subseteq D_{k-1}(F)$ of $\beta(a)$ with $U_0 \ll \beta(U)$. Clearly $f(U_0)$ is a join-cover of a and $f(U_0) \ll U$. Furthermore, the set $\beta f(U_0)$ is a join-cover of $\beta(a)$ with $\beta f(U_0) \ll U_0$. By the minimality of U_0, we have $U_0 \subseteq \beta f(U_0)$, and since U_0 is finite, this implies $\beta f(U_0) = U_0 \subseteq D_{k-1}(F)$. It now follows from the induction hypothesis that $f(U_0) \subseteq D_{k-1}$, and therefore $a \in D_k(L)$.

Now assume $D(L) = L$, and consider a homomorphism $f : K \to L$ where K is generated by a finite set Y. Let $H_0 = Y$, and for $k \in \omega$ let H_{k+1} be the set of all joins of meets of elements in H_k. For each $k \in \omega$ define maps $\beta_k : L \to K$ by

$$\beta_k(a) = \prod\{y \in H_k : f(y) \ge a\}.$$

(In particular $\prod \emptyset = \sum Y = 1_K$.) We claim that for every $k \in \omega$ and $a \in D_k(L)$, if the set $f^{-1}[a)$ is nonempty, then $f^{-1}[a) = [\beta_k(a))$, and $\beta_k(a)$ is therefore the smallest element of this set. Since $D(L) = L$ it then follows that f is lower bounded. So suppose that $a \in D_k(L)$ for some k, and $f^{-1}[a)$ is nonempty. If $x \in K$ and $x \geq \beta_k(a)$ then

$$ f(x) \geq f\beta_k(a) = \begin{cases} f(1_K) & \text{if } \{y \in H_k : f(y) \geq a\} = \emptyset \\ \prod\{f(y) : y \in H_k \text{ and } f(y) \geq a\} & \text{otherwise,} \end{cases} $$

and in both cases $f(x) \geq a$ ($f(1_K) \geq a$ since $f^{-1}[a)$ is nonempty). Thus $[\beta_k(a)) \subseteq f^{-1}[a)$. For the reverse inclusion we have to show that for all $x \in K$

$$ (*) \qquad f(x) \geq a \qquad \text{implies} \qquad x \geq \beta_k(a). $$

The set of elements $x \in K$ that satisfy $(*)$ contains Y and is closed under meets, hence it is enough to show that it is also closed under joins. For $k = 0$, we have $a \in D_0(L)$, so $\sum f(U) \geq a$ implies $f(u) \geq a$ for some $u \in U$, and by $(*)$ $u \geq \beta_0(a)$, hence $\sum U \geq \beta_0(a)$. Suppose now that $(*)$ holds for all values less than some fixed $k > 0$. Let $x = \sum U$ and assume $f(x) \geq a$, i.e. $f(U)$ is a join-cover of a. If it is trivial, then $(*)$ is satisfied as before, so assume it is nontrivial. Then there exists a join-cover $V \subseteq D_{k-1}(L)$ of a with $V \ll f(U)$, and by the inductive hypothesis $x \geq \beta_{k-1}(v)$ for all $v \in V$. Now the elements $\beta_{k-1}(v)$ are meets of elements in H_{k_1}, and the element $z = \sum \beta_{k-1}(V)$ therefore belongs to H_k. Since $f(z) \geq \sum V \geq a$, it follows from the definition of β_k that $z \geq \beta_k(a)$, hence $x \geq \beta_k(a)$ as required. (iii) implies (i) follows immediately from the assumption that L is finitely generated. $\qquad \square$

The equivalence of (i) and (iii) was originally proved by McKenzie [**72**]. Note that (i)$\Rightarrow$(ii)$\Rightarrow$(iii) is true for any lattice L, so the above theorem implies that if L is lower bounded then $D(L) = L$, and the converse holds whenever L is finitely generated. Together with Lemma 2.21 we also have that every finitely generated sublattice of a free lattice is bounded.

It is fairly easy to compute $D(L)$ and $D'(L)$ for any given finite lattice L. Thus one can check that the lattices N, L_6, $L_7, \ldots, L_{15}$ are all bounded, and since they also satisfy Whitman's condition (W), Theorem 2.19 implies that they are projective (hence sublattices of a free lattice). On the other hand L_1 fails to be upper bounded (dually for L_2) and M_3, L_3, L_4 and L_5 are neither upper nor lower bounded (see Figures 2.1 and 2.2).

Note that if U is a finite subset of $D_k(L)$ then $\sum U \in D_{k+1}(L)$. Since every join irreducible element of a distributive lattice is join prime, and dually, it follows that every finite distributive lattice is bounded.

We now recall a construction which is usually used to prove that the variety of all lattices is generated by its finite members. In Section 1.2 it was shown that every free lattice $F(X)$ can be constructed as a quotient algebra of a word algebra $W(X)$, whence elements of $F(X)$ are represented by lattice terms (words) of $W(X)$. The *length* λ of a lattice term is defined inductively by $\lambda(x) = 1$ for each $x \in X$ and $\lambda(p + q) = \lambda(pq) = \lambda(p) + \lambda(q)$ for any terms $p, q \in W(X)$. Let X be a finite set, and for each $k \in \omega$, construct a finite lattice $P(X, k)$ as follows:

Take W to be the finite subset of the free lattice $F(X)$ which contains all elements that can be represented by lattice terms of length at most k, and let $P(X, k)$ be the set of all finite meets of elements from W, together with largest element $1_F = \sum X$ of $F(X)$. Then $P(X,k)$ is a finite subset of $F(X)$, and it is a lattice under the partial order inherited from $F(X)$, since it is closed under meets and has a largest element. However, $P(X,k)$ is not a sublattice of $F(X)$ because for $a, b \in P(X,k)$, $a +_P b \geq a +_F b$ and equality holds if and only if $a +_F b \in P(X,k)$. Nevertheless, $P(X,k)$ is clearly generated by the set X.

LEMMA 2.24

(i) If $p = q$ is a lattice identity that fails in some lattice, then $p = q$ fails in a finite lattice of the form $P(X,k)$ from some finite set X and $k \in \omega$.

(ii) If $h : F(X) \twoheadrightarrow P(X,k)$ is the extension of the identity map on X, then h is upper bounded.

PROOF. (i) Let X be the set of variables that occur in p and q, and let k be the greater of the lengths of p and q. Since $p = q$ fails in some lattice, p and q represent different elements of the free lattice $F(X)$ and therefore also different elements of $P(X,k)$. Thus $p = q$ fails in $P(X,k)$. (ii) Note that for $a \in P(X,k) \subseteq F(X)$, $h(a) = a$, and in general $h(b) \geq_F b$ for any $b \in F$. Therefore $h(b) \leq_P a$ implies $b \leq_F h(b) \leq_P a$, and conversely $b \leq_F a$ implies $h(b) \leq_P h(a) = a$, whence a is the largest element of $f^{-1}(a]$ for all $a \in P(X,k)$. $\square$

THEOREM 2.25 (McKenzie [72]). S is a splitting lattice if and only if S is a finite subdirectly irreducible bounded lattice

PROOF. Suppose S is a splitting lattice, and $p = q$ is its conjugate identity. We have to show that S is a bounded epimorphic image of some free lattice $F(X)$. As we noted in the beginning of Section 2.3, every splitting lattice is finite, so there exists an epimorphism $h : F(X) \twoheadrightarrow S$ for some finite set X. The identity $p = q$ does not hold in S, hence it fails in $F(X)$ and also in $P(X,k)$ for some large enough $k \in \omega$ (by Lemma 2.24 (i)). Therefore $S \in \{P(X,k)\}^V$ and since S is subdirectly irreducible and $P(X,k)$ is finite, it follows from Jónsson's Lemma (Corollary 1.7(i)) that $S \in \mathbf{HS}\{P(X,k)\}$. So there exists a sublattice L of $P(X,k)$ and an epimorphism $g : L \twoheadrightarrow S$. Since g is onto, we can choose for each $x \in X$ an element $a_x \in L$ such that $g(a_x) = h(x)$. Let $f : F(X) \to L$ be the extension of the map $x \mapsto a_x$. By Lemma 2.24 (ii) $P(X,k)$ is an upper bounded image of $F(X)$, hence by the equivalence of (i) and (iii) of (the dual of) Theorem 2.23 f is upper bounded. Since L is finite, g is obviously bounded, and therefore $h = gf$ is upper bounded. A dual argument shows that h is also lower bounded, whence S is a bounded lattice.

Conversely, suppose S is a finite subdirectly irreducible lattice, and let u/v be a prime critical quotient of S. If S is bounded, then there exists a bounded epimorphism h from some free lattice $F(X)$ onto S. Let r be the smallest element of $h^{-1}[u)$ and let s be the largest element of $h^{-1}(v]$. Now $r + s/s$ is a prime quotient of $F(X)$, for if $s < t \leq r + s$ then $h(t) = u = h(tr) = h(r)$, and by the choice of r, $tr = r$ hence $r \leq t = r + s$. By Lemma 1.10 there exists a largest congruence θ on $F(X)$ which does not identify $r + s$ and s. Since $h(r + s) = u \neq v = h(s)$ we have $\ker h \subseteq \theta$, and equality follows from the fact that u/v is a critical quotient of S. Now Corollary 2.12 implies that S is a splitting lattice. $\square$

Referring to the remark after Theorem 2.23 we note that the lattices $L_6, L_7, \ldots, L_{15}$ are examples of splitting lattices. In fact McKenzie [72] shows how one can effectively compute a conjugate identity for any such lattice. For the details of this procedure we refer the reader to his paper and also to the more recent work of Freese and Nation [83].

From Corollary 2.12 and the proof of the above theorem we obtain the following:

COROLLARY 2.26 (McKenzie [72]). *A lattice L is a splitting lattice if and only if L is isomorphic to $F(n)/\psi_{rs}$ where $F(n)$ is some finitely generated free lattice and ψ_{rs} is the largest congruence that does not identify some covering pair $r \succ s$ of $F(n)$.*

Canonical representations and semidistributivity. A finite set U of a lattice L is said to be a *join representation* of an element a in L if $a = \sum U$. Thus a join representation is a special case of a join-cover. U is a *canonical* join representation of a if it is irredundant (i.e. no proper subset of U is a join representation of a) and refines every other join representation of a. Note that an element can have at most one canonical join representation, since if U and V are both canonical join representations of a then $U \ll V \ll U$ and because the elements of an irredundant join representation are noncomparable it follows that $U \subseteq V \subseteq U$. However canonical join representations do not exist in general (consider for example the largest element of M_3).

Canonical meet representations are defined dually and have the same uniqueness property.

A fundamental result of Whitman's [41] paper is that every element of a free lattice has a canonical join representation and a canonical meet representation. We briefly outline the proof of this result. Denote by $\bar{p}$ the element of $F(X)$ represented by the term p. A term p is said to be *minimal* if the length of p is minimal with respect to the lengths of all terms that represent $\bar{p}$. If p is formally a join of simpler terms $p_1, \ldots, p_n$, none of which is itself a join, then these terms will be called the *join components* of p. The *meet components* of p are defined dually. Note that every term is either a variable or it has join components or meet components.

THEOREM 2.27 (Whitman[41]). *A term p is minimal if and only if $p = x \in X$ or p has join components $p_1, \ldots, p_n$ and for each $i = 1, \ldots, n$*

(1) *p_i is minimal,*

(2) *$\bar{p}_i \not\leq \sum_{j \neq i} \bar{p}_j$,*

(3) *for any meet component r of p_i, $\bar{r} \neq \bar{p}$,*

or the duals of (1), (2) and (3) hold for the meet components of p.

PROOF. All $x \in X$ are minimal, so by duality we may assume that p has join components $p_1, \ldots, p_n$. If (1), (2) or (3) fail, then we can easily construct a term q such that $\bar{q} = \bar{p}$, but $\lambda(q)\lambda(p)$, which shows that p is not minimal. (If (1) fails, replace a nonminimal p_i by a minimal term; if (2) fails, omit the p_i for which $\bar{p}_i \leq \sum_{j \neq i} \bar{p}_j$; if (3) fails, replace p_i by its meet component r which satisfies $\bar{r} \leq \bar{p}$.)

Conversely, suppose p satisfies (1), (2) and (3), and let q be a minimal term such that $\bar{q} = \bar{p}$. We want to show that $\lambda(p) = \lambda(q)$, then p is also minimal. First observe that q must have join components, for if $q \in X$ or if q has meet components $q_1, \ldots, q_m$, then $\bar{q} \leq \bar{p}_1 + \ldots + \bar{p}_n$ together with (W3) or (W) imply $\bar{q} \leq \bar{p}_i \leq \bar{p}$ or $\bar{q} \leq \bar{q}_j \leq \bar{p}$ for some i, and since $\bar{q} = \bar{p}$ we must have equality throughout, which contradicts the minimality of q.

So let $q_1, \ldots, q_m$ be the join components of q. For each $i \in \{1, \ldots, n\}, \bar{p}_i \leq \bar{q}_1 + \ldots + \bar{q}_m$ and either $p_i \in X$ or p_i has meet components (whose images in $F(X)$ are not below $\bar{p}$ by condition (3)), so we can use (W3) or (W) to conclude that $\bar{p}_i \leq \bar{q}_{i^*}$ for some unique $i^* \in \{1, \ldots, m\}$. Similarly, since q is minimal, it satisfies (1)–(3), and so for each $j \in \{1, \ldots, m\}$ there exists a unique $j_* \in \{1, \ldots, n\}$ such that $\bar{q}_j \leq \bar{p}_{j_*}$. Thus $\bar{p}_i \leq \bar{q}_{i^*} \leq \bar{p}_{i^*_*}$ and $\bar{q}_j \leq \bar{p}_{j_*} \leq \bar{q}_{j_*^*}$, whence (2) implies $i = i^*_*$ and $j = j_*^*$. It follows that the map $i \mapsto i^*$ is a bijection, $m = n$, $\bar{p}_i = \bar{q}_{i^*}$ and since both terms are minimal by (1), $\lambda(p_i) = \lambda(q_{i^*})$. Consequently $\lambda(p) = \lambda(q)$.   □

COROLLARY 2.28 *Every element of a free lattice has a canonical join representation and a canonical meet representation.*

PROOF. Suppose u is an element of a free lattice, and let p be a minimal term such that $\bar{p} = u$. If p has no join components, then (W3) or (W) imply that u is join irreducible, in which case $\{u\}$ is the canonical join representation of u. If p has join components $p_1, \ldots, p_n$ then condition (2) above implies that $U = \{\bar{p}_1, \ldots, \bar{p}_n\}$ is irredundant, and (W3) or (W) and condition (3) imply that U is a canonical join representation of u. The canonical meet representation is constructed dually.   □

The existence of canonical representations is closely connected to the following weak form of distributivity:

A lattice L is said to be *semidistributive* if it satisfies the following two implications for all $u, x, y, z \in L$:

$$(SD^+) \qquad u = x + y = x + z \quad \text{implies} \quad u = x + yz \qquad \text{and dually}$$
$$(SD^\cdot) \qquad u = xy = xz \qquad \text{implies} \quad u = x(y + z).$$

LEMMA 2.29 *If every element of a lattice L has a canonical join representation then L satisfies (SD^+).*

PROOF. Let $u = \sum V$ be a canonical join representation of u, and suppose $u = x + y = x + z$. Then for each $v \in V$ we have $v \leq x$ or $v \leq y, z$. It follows that $v \leq x + yz$ for each $v \in V$, which implies $u \leq x + yz$. The reverse inclusion always holds, hence (SD^+) is satisfied.   □

Now Corollary 2.28 and the preceding lemma together with its dual imply that *every free lattice is semidistributive.* The next lemma extends this observation to all bounded lattices.

LEMMA 2.30

(i) *Bounded epimorphisms preserve semidistributivity.*

(ii) *Every bounded lattice is semidistributive.*

PROOF. (i) Suppose L is semidistributive and $f : L \twoheadrightarrow L'$ is a bounded epimorphism. Let $u, x, y, z \in L'$ be such that $u = x + y = x + z$. Then $\beta(u) = \beta(x) + \beta(y) = \beta(x) + \beta(z) = \beta(x) + \beta(y)\beta(z)$, where $\beta = \beta_f : L' \hookrightarrow L$ is the join-preserving map associated with f. Hence

$$u = f\beta(u) = f(\beta(x) + \beta(y)\beta(z)) = x + yz,$$

which shows that (SD$^+$) holds in L'. (SD$^-$) follows by duality (using α_f), whence L' is semidistributive. Now (ii) follows immediately from the fact that every free lattice is semidistributive.   □

Since there are lattices in which the semidistributive laws fail (the simplest one is the diamond M_3) it is now clear that these lattices cannot be bounded. However, there are also semidistributive lattices which are not bounded. An example of such a lattice is given at the end of this section (Figure 2.2).

For finite lattices the converse of Lemma 2.29 also holds. To see this we need the following equivalent form of the semidistributive laws.

LEMMA 2.31 (Jónsson and Kiefer [62]). *A lattice* L *satisfies* (SD$^+$) *if and only if for all* $u, a_1, \ldots, a_m, b_1, \ldots, b_n \in L$

$$(*) \qquad u = \sum_{i=1}^m a_i = \sum_{j=1}^n b_j \qquad implies \qquad u = \sum_{i=1}^m \sum_{j=1}^n a_i b_j.$$

PROOF. Assuming that L satisfies (SD$^+$), we will prove by induction that the statement

$$P(m,n) \qquad \begin{array}{c} \text{for all} \quad w, a_1, \ldots, a_m, b_1, \ldots, b_n \in L \\ u = w + \sum_i a_i = w + \sum_j b_j \qquad \text{implies} \qquad u = w + \sum_i \sum_j a_i b_j \end{array}$$

holds for all $m, n \geq 1$. Then $(*)$ follows if we choose any $w \leq \sum_i \sum_j a_i b_j$.

$P(1,1)$ is precisely (SD$^+$), so we assume that $n > 1$, and that $P(1, n')$ holds whenever $1 \leq n' < n$. By hypothesis $u = w + a_1 = w + b_n' + b_n$, where $b_n' = \sum_{j=1}^{n-1} b_j$. Therefore

$$u = (w + b_n') + a_1 = (w + b_n') + b_n \qquad \text{implies}$$
$$u = w + b_n' + a_1 b_n \qquad \text{by (SD}^+\text{), and now}$$
$$u = w + a_1 b_n + a_1 = w + a_1 b_n + \sum_{j=1}^{n-1} b_j \qquad \text{implies}$$
$$u = w + a_1 b_n + \sum_{j=1}^{n-1} a_1 b_j = w + \sum_{j=1}^n a_1 b_j \qquad \text{by } P(1, n-1).$$

Hence $P(1, n)$ holds for all n.

Now assume that $m > 1$ and that $P(m', n)$ holds for $1 \leq m' < m$. By hypothesis $u = w + a_m' + a_m = w + \sum_{j=1}^n b_j$, where $a_m' = \sum_{i=1}^{m-1} a_i$. Consequently

$$u = w + a_m' + a_m = w + a_m' + \sum_{j=1}^n b_j \qquad \text{implies}$$
$$u = w + a_m' + \sum_{j=1}^n a_m b_j \qquad \text{by } P(1, n), \text{ and now}$$
$$u = (w + \sum_{j=1}^n a_m b_j) + \sum_{i=1}^m a_i = (w + \sum_{j=1}^n a_m b_j) + \sum_{j=1}^n b_j \qquad \text{implies}$$
$$u = w + \sum_{j=1}^n a_m b_j + \sum_{i=1}^{m-1} \sum_{j=1}^n a_i b_j = w + \sum_{i=1}^m \sum_{j=1}^n a_i b_j$$

by $P(m-1, n)$ as required. Therefore $P(m, n)$ holds for all m, n.

Conversely, if $(*)$ holds and $u = a + b = a + c$ for some $u, a, b, c \in L$, then $u = aa + ab + ac + bc = a + bc$. Hence (SD$^+$) holds in L.   □

COROLLARY 2.32 (Jónsson and Kiefer [62]). *A finite lattice satisfies* (SD$^+$) *if and only if every element has a canonical join representation.*

PROOF. Let L be a finite lattice that satisfies (SD$^+$), and suppose that V and W are two join representations of $u \in L$. By the preceding lemma the set $\{ab : a \in V, \ b \in W\}$ is again a join representation of u, and it clearly refines both V and W. Since L is finite, u has only finitely many distinct join representations. Combining these in the same way we obtain a join representation U that refines every other join representation of u. Clearly a canonical join representation is given by a subset of U which is an irredundant join representation of u. The converse follows from Lemma 2.29   □

Thus finite semidistributive lattices have the same property as free lattices in the sense that every element has canonical join and meet representations. Further results about semidistributivity appear in Section 4.2.

Cycles in semidistributive lattices. We shall now discuss another way of characterizing splitting lattices, due to Jónsson and Nation [75]. Let L be a finite lattice and denote by $J(L)$ the set of all nonzero join-irreducible elements of L. Every element $p \in J(L)$ has a unique lower cover, which we denote by p_*. We define two binary relations A and B on the set $J(L)$ as follows: for $p, q \in J(L)$ we write

$$pAq \quad \text{if} \quad p < q + x, \ q < p, \ q \not\leq x \text{ and } q_* \leq x \text{ for some } x \in L$$
$$pBq \quad \text{if} \quad p \leq p_* + q, \ p \not\leq q \text{ and } p \not\leq p_* + q_*.$$

A third relation σ is defined by $p\sigma q$ if pAq or pBq. Note that if pAq then $qx = q_*$, $p + x = q + x$ and $px \geq q_*$. So, depending on whether or not the last inequality is strict, the elements p, q, x generate a sublattice of L that is isomorphic to either A_1 or A_2 (Figure 2.6). Also if pBq, then $p_* \not\leq q_*$ (else $p \leq p_* + q \leq q_* + q = q$, contradicting $p \not\leq q$) and $p + q = p_* + q \geq p + q_*$. Now the elements p, p_*, q, q_* generate a sublattice of L isomorphic to

$$
\begin{array}{ll}
B_1 & \text{if} \quad q_* \not\leq p_* \text{ and } p_* + q > p + q_* \\
B_2 & \text{if} \quad q_* \leq p_* \\
B_3 & \text{if} \quad q_* \not\leq p_* \text{ and } p_* + q = p + q_*.
\end{array}
$$

If we assume that L is semidistributive, then the last case is excluded since B_3 fails (SD^+) $(p_* + q_* + p = p_* + q_* + q \neq p_* + q_* + pq)$. Observe also that in general the element x in the definition of A is not unique, but in the presence of (SD$^.$) we can always take $x = \kappa(q)$, where

$$\kappa(q) = \sum \{ x \in L : q \not\leq x \text{ and } q_* \leq x \} = \sum \{ x \in L : qx = q_* \},$$

since L is finite and by (SD$^.$) $\kappa(q)$ itself satisfies $q\kappa(q) = q_*$. In this case x is covered by $q + x$.

The following lemma from Jónsson and Nation [**75**] motivates the above definitions. Note that if $D(L) \neq L$ for some finite lattice L, then some join-irreducible element of L is not in $D(L)$, since for any nonempty subset U of $D_k(L)$, $\sum U$ is an element of $D_{k+1}(L)$.

LEMMA 2.33 *If L is a finite semidistributive lattice and $p \in J(L) - D(L)$ then there exists $q \in J(L) - D(L)$ with $p\sigma q$.*

PROOF. Since $p \notin D(L)$, there exists a nontrivial join-cover V of p such that no join-cover $U \subseteq D(L)$ of p refines V. Since $p \leq \sum V$, we have $\sum V \not\leq \kappa(p)$, whence $v_0 \not\leq \kappa(P)$ for some $v_0 \in V$. Choose $y \leq v_0$ minimal with respect to the property $y \not\leq k(p)$. Clearly $y \in J(L)$ and $p \leq y$ since V is a nontrivial join-cover. Note that $y \not\leq \kappa(p)$ if and only if $p \leq p_* + y$, so by the minimality of y, $p \not\leq p_* + y_*$. Thus pBy, and if $y \notin D(L)$, then $q = y$ yields the desired conclusion.

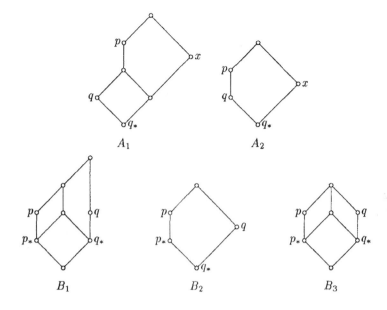

Figure 2.6

Otherwise $y \in D(L)$, and we can choose an element $z \leq p_*$ subject to the condition $p < y + z$. We claim that $z \notin D(L)$. Assume the contrary. Since $z < p \leq \sum V$, V is a join-cover of z, which is either trivial or nontrivial. If it is trivial, we let $U = \{y, z\}$, and if it is nontrivial, then there exists a join-cover $W \subseteq D(L)$ of z which refines V, and we let $U = W \cup \{y\}$. In both cases U is a subset of $D(L)$ and a join-cover of p which refines V (since $y \leq v_0$), contradicting the assumption $p \notin D(L)$.

By Corollary 2.32 every element of L has a canonical join representation, so there exists a finite set $U_0 \subseteq L$ such that $z = \sum U_0$. Since $z \notin D(L)$, there exists $q \in U_0$ such that $q \notin D(L)$. Letting $q' = \sum (U_0 - \{q\})$ and $x = y + q_* + q'$ we see that $x + q \geq y + z > p$, $q_* \leq x$ and $q < p$ (since $z \leq p_*$). Furthermore, $q_* + q' < z$ and therefore $p \not\leq x$ by the minimality of z. Consequently pAq. $\qquad\square$

By a repeated application of the preceding lemma we obtain elements $p_i \in J(L)$ with $p_i \sigma p_{i+1}$ for $i \in \omega$. Since L is finite, this sequence must repeat itself eventually, so we can assume that $p_0 \sigma p_1 \sigma \cdots \sigma p_n \sigma p_0$. Such a sequence is called a *cycle*, and it follows that the nonexistence of such cycles in a finite semidistributive lattice L implies that $D(L) = L$. The next result, also from Jónsson and Nation [75], shows that the converse is true is an arbitrary lattice.

THEOREM 2.34 *If L is any lattice which contains a cycle, then $D(L) \neq L$.*

PROOF. Suppose $p_0 \sigma p_1 \sigma \cdots \sigma p_n \sigma p_0$ for some $p_i \in J(L)$. From Figure 2.6 we can see that each p_i has a nontrivial cover, so $p_i \notin D_0(L)$. Suppose no p_i belongs to $D_{k-1}(L)$, but say $p_0 \in D_k(L)$. If $p_0 A p_1$, then $p_0 < p_1 + x$, $p_1 < p_0$, $p_1 \not\leq x$ and $p_{1*} \leq x$ for some $x \in L$.

Since $\{p_1, x\}$ covers p_0, there exists a cover $U \subseteq D_{k-1}(L)$ of p_0 with $U \ll \{p_1, x\}$. For each $u \in U$, either $u \leq x$ or $u < p_1$ (since $p_1 \notin D_{k-1}(L)$), but if $u < p_1$, then $u \leq p_{1*} \leq x$. Hence $u \leq x$ for all $u \in U$, so that $p_0 \leq \sum U \leq x$, a contradiction.

If $p_0 B p_1$, then $\{p_{0*}, p_1\}$ covers p_0, and therefore there exists a cover $u \subseteq D_{k-1}(L)$ of p_0 with $U \ll \{p_{0*}, p_1\}$. For each $u \in U$, either $u \leq p_{0*}$ or $u < p_1$, i.e. $u \leq p_{0*}$ or $u \leq p_{1*}$. Therefore $p_0 \leq \sum U \leq p_{0*} + p_{1*}$, again a contradiction. By induction we have $p_i \notin D(L)$ for all i, and so $D(L) \neq L$.   □

COROLLARY 2.35 *For a finite semidistributive lattice* L, $D(L) = L$ *if and only if* L *contains no cycles.*

An example of a finite semidistributive lattice which contains a cycle is give in Figure 2.7. at the end of this section.

Day's characterization of finite bounded lattices. The results of this section are essentially due to Alan Day, but the presentation here is taken from Jónsson and Nation [**75**]. A more general treatment can be found in Day [**79**].

We investigate the relationship between $J(L)$ and $J(\text{Con}(L))$. By transitivity, a congruence relation on a finite lattice is determined uniquely by the prime quotients which it collapses. The next lemma shows that we need only consider prime quotients of the form p/p_*, where $p \in J(L)$.

LEMMA 2.36 *Let* L *be a finite lattice, and suppose* $\theta \in \text{Con}(L)$. *Then*

(i) $\theta \in J(\text{Con}(L))$ *if and only if* $\theta = \text{con}(u, v)$ *for some prime quotient* u/v *of* L;

(ii) *if* u/v *is a prime quotient of* L *then there exists* $p \in J(L)$ *such that* $p/p_* \nearrow u/v$;

(iii) *if* L *is semidistributive, then the element* p *in* (ii) *is unique.*

PROOF. (i) In a finite lattice

$$\theta = \sum \{\text{con}(u, v) : u/v \text{ is prime and } u\theta v\}$$

so if θ is join-irreducible then $\theta = \text{con}(u, v)$ for some prime quotient u/v. Conversely, suppose $\phi \in \text{Con}(L)$ is strictly below θ. Then $\phi \subseteq \psi_{uv} \cap \text{con}(u, v)$, where ψ_{uv} is the unique largest congruence that does not identify u and v. Hence $\psi_{uv} \cap \text{con}(u, v)$ is the unique dual cover of $\text{con}(u, v)$, and it follows that $\text{con}(u, v) \in J(\text{con}(L))$. To prove (ii) we simply choose p minimal with respect to the condition $u = v + p$. Then $p \in J(L)$ and $p/p_* \nearrow u/v$.

(iii) Suppose for some $q \in J(L)$, $q \neq p$, we also have $q/q_* \nearrow u/v$. Then $u = v + p = v + q$, and by semidistributivity $u = v + pq$. Now $p \neq q$ implies $pq < q$ or $pq < p$, so $pq \leq p_*$ or $pq \leq q_*$. But then $pq \leq v$, hence $u = v + pq = v$, which is a contradiction.   □

From (i) and (ii) we conclude that for any finite lattice L the map $p \mapsto \text{con}(p, p_*)$ from $J(L)$ to $J(\text{Con}(L))$ is onto. Day [**79**] shows that the map is one-one if and only if L is lower bounded. Let us say that a set Q of prime quotients in L corresponds to a congruence relation θ on L if θ collapses precisely those prime quotients in L that belong to Q.

LEMMA 2.37 *A set of prime quotients Q in a finite lattice L corresponds to some $\theta \in$ Con(L) if and only if*

(∗) *For any two prime quotients r/s and u/v in L, if $r/s \in Q$ and if either $s \le v < u \le r + v$ or $su \le v < u \le r$, then $u/v \in Q$.*

PROOF. The two conditions of (∗) can be rewritten as either $r/s \nearrow r + v/v \supseteq u/v$ or $r/s \searrow u/su \supseteq u/v$, hence if r/s is collapsed by some congruence, so is u/v. Therefore (∗) is clearly necessary.

Suppose (∗) holds and let $\theta = \sum \{\mathrm{con}(u,v) : u/v \in Q\}$. If x/y is a prime quotient that is collapsed by θ, then $\mathrm{con}(x,y) \subseteq \theta$. But $\mathrm{con}(x,y)$ is join irreducible and Con(L) is distributive, so $\mathrm{con}(x,y)$ is in fact join prime, whence $\mathrm{con}(x,y) \subseteq \mathrm{con}(u,v)$ for some $u/v \in Q$. By Lemma 1.11 u/v projects weakly onto x/y, and since (∗) forces each quotient in the sequence of transposes to be in Q, it follows that $x/y \in Q$. □

THEOREM 2.38 (Jónsson and Nation [75]). *If L is a finite semidistributive lattice and $S \subseteq J(L)$, then the following conditions are equivalent:*

(i) *There exists $\theta \in$ Con(L) such that for all $p \in J(L)$, $p\theta p_*$ if and only if $p \in S$.*

(ii) *For all $p, q \in J(L)$, $p\sigma q$ and $q \in S$ imply $p \in S$.*

PROOF. Assume (i) and let $p, q \in J(L)$ with $p\sigma q$ and $q \in S$. Then Figure 2.6 shows that q/q_* projects weakly onto p/p_*, so if θ collapses q/q_*, it also collapses p/p_*, which implies $p \in S$.

Conversely, suppose (ii) holds, and let Q be the set of all prime quotients u/v in L, such that the unique member $p \in J(L)$ with $p/p_* \nearrow u/v$ belongs to S. Then $p/p_* \in Q$ if and only if $p \in S$, so by the proceeding lemma it suffices to show that Q satisfies (∗). Consider two prime quotients r/s and u/v in L and let p and q be the corresponding members of $J(L)$, so that $q/q_* \nearrow r/s$ and $p/p_* \nearrow u/v$. If $r/s \in Q$, then by uniqueness $q \in S$. Now $p = q \in S$ implies $u/v \in Q$ by definition. Assuming that $p \ne q$, we are going to show that

(1) if $s \le v < u \le r + v$ then pBq, and

(2) if $su \le v < u \le r$ then pAq.

Statement (ii) then implies $p \in S$, whence $u/v \in Q$ as required. Under the hypothesis of (1) we need to show that $p \le p_* + q$, $p \not\le q$ and $p \not\le p_* + q_*$. Since $q_* \le s \le v$ and $p_* \le v$, we must have $p \not\le p_* + q_*$, else $p \le v$. For the same reason $p \not\le q$, and $p \ne q$ by assumption. Finally, $p \not\le p_* + q$ would imply $p(p_* q) = p_*$, which together with $pv = p_*$ gives $p(v+q) = p_*$ by semidistributivity. This, however, is impossible since $v + q = s + v + q = r + v \ge u \ge p$. This proves (1).

Now suppose that the hypothesis of (2) is satisfied. Clearly $q \not\le s$ and $q_* \le s$, so to prove pAq it suffices to show that $q < p$ and $p < q + s$. We certainly have $q + s = r \ge p$ by the hypothesis, and this inclusion must be strict, since $p = r > s$ would imply $p = q$ by the join irreducibility of p. Observe that $p \not\le s$ because $ps \le su \le v$ and $p \not\le v$. Since $p \le r$, this implies $r = s + p$ which, together with $r = s + q$ yields $r = s + pq$. Now $q \not\le p$ would imply $pq \le q_* \le s$, which is impossible because $s + pq = r > s$. Thus $p \le q$ as required. □

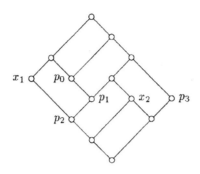

Figure 2.7

THEOREM 2.39 *For any finite semidistributive lattice L, the following conditions are equivalent:*

  (i) $|J(L)| = |J(\mathrm{Con}(L))|$

  (ii) $D(L) = L$

  (iii) $D'(L) = L$

  (iv) L *is bounded.*

PROOF. It follows from the preceding theorem that, for two distinct elements p and q of $J(L)$, $\mathrm{con}(p, p_*) = \mathrm{con}(q, q_*)$ if and only if there exists a cycle containing both p and q. Consequently the map $p \mapsto \mathrm{con}(p, p_*)$ from $J(L)$ to $J(\mathrm{Con}(L))$ is one-one if and only if L contains no cycles if and only if $D(L) = L$ by Corollary 2.35. Therefore (i) is equivalent to (ii).

    Lemma 2.36(iii) and its dual imply that the number of meet irreducible elements of L is equal to the number of join irreducible elements (to every prime quotient m^*/m where m is meet irreducible and m^* is its dual cover, corresponds a unique $p \in J(L)$ such that $p/p_* \nearrow m^*/m$, and vice versa). Therefore the condition $|J(L)| = |J(\mathrm{Con}(L))|$ is equivalent to its own dual, and hence to the dual of $D(L) = L$, namely $D'(L) = L$.

    Lastly (ii) and (iii) together are equivalent to (iv) by Theorem 2.23 and its dual. □

    It is interesting to examine how the above conditions fail in the semidistributive lattice in Figure 2.7, which contains the cycle $p_0 \, A \, p_1 \, A \, p_2 \, B \, p_3 \, B \, p_0$. If we add an element a on the edge c, p_2, then we obtain an example of a subdirectly irreducible semidistributive lattice which is not a splitting lattice. It is not difficult to prove that every critical quotient of a splitting lattice must be prime, but the example we just mentioned shows that the converse does not hold even for semidistributive lattices.

2.4 Splitting lattices generate all lattices

We will now prove a few lemmas which lead up to the result of Day [77], that the variety
$\mathcal{L}$ of all lattices is generated by the class of all splitting lattices. This result and some of
the characterizations of splitting lattices will be used at the end of Chapter 6.

Let $\mathcal{B}$ be the class of all bounded lattices and let $\mathcal{B}_F$ be the class of finite members of
$\mathcal{B}$. By Theorem 2.25 $(\mathcal{B}_F)_{SI}$ is the class of all splitting lattices, and it is clearly sufficient
to show that $\mathcal{L} = (\mathcal{B}_F)^{\mathcal{V}}$.

LEMMA 2.40 $\mathcal{B}_F$ *is closed under sublattices, homomorphic images and direct products
with finitely many factors.*

PROOF. If L is a sublattice of a lattice $B \in \mathcal{B}_F$, then by Lemma 2.17 (iv), any $f : F(X) \to
\to L$, where $F(X)$ is a finitely generated free lattice, is bounded. If L is a homomorphic
image of B, say $h : B \twoheadrightarrow L$, then there exists $g : F(X) \to B$ such that $f = hg$, and
by the equivalence of (i) and (iii) of Theorem 2.23 g is bounded. h is bounded since B
is finite, hence f is also bounded. Lastly, if $B_1, B_2 \in \mathcal{B}_F$ and $f : F(X) \twoheadrightarrow B_1 \times B_2$ is
an epimorphism, then $\pi_i f$ is bounded, where $\pi_i : B_1 \times B_2 \twoheadrightarrow B_i$ is the projection map
$(i = 1, 2)$. For $b_i \in B_i$, let $\beta_i(b_i)$ be the least preimage of b_i under the map $\pi_i f$, and
denote the zero of B_i by 0_i. Since $(b_1, 0_2)$ is the least element of $\pi_1^{-1}\{b_1\}$, $\beta_1(b_1)$ is also the
least element of $f^{-1}\{(b_1, 0_2)\}$, and similarly $\beta_2(b_2)$ is the least element of $f^{-1}\{(0_1, b_2)\}$.
It follows that $\beta_1(b_1) + \beta_2(b_2)$ is the least preimage of $(b_1, 0_2) + (0_1, b_2) = (b_1, b_2)$ under f.
Hence f is lower bounded, and a dual argument shows that f is also upper bounded. □

The two element chain is a splitting lattice, so the above lemma implies that every
finite distributive lattice is bounded. Recall the construction of the lattice $L[u/v]$ from a
lattice L and a quotient u/v of L (see above Lemma 2.15).

LEMMA 2.41 (Day [77]). *If $L \in \mathcal{B}_F$ and $I = u/v$ is a quotient of L, then $L[I] \in \mathcal{B}_F$.*

PROOF. By assumption L is a finite lattice, hence $L[I]$ is also finite. Let X be a finite
set with $f : F(X) \twoheadrightarrow L[I]$ a lattice epimorphism and let $\gamma : L[I] \twoheadrightarrow L$ be the natural
epimorphism. Since $L \in \mathcal{B}_F$, $h = \gamma f : F(X) \twoheadrightarrow L$ is bounded, so for each $b \in L$ there
exists a least member $\beta_h(b)$ of $h^{-1}\{b\}$. By definition of γ, we have

$$\gamma^{-1}\{b\} = \begin{cases} \{b\} & \text{if } b \in L - I \\ \{(b,0),(b,1)\} & \text{if } b \in I. \end{cases}$$

hence $\beta_h\gamma(b)$ is the least member of $f^{-1}\{b\}$ for each $b \in (L - I) \cup I \times \{0\}$. Note that for
any $a_1, a_2, b_1, b_2 \in L$ if a_i is the least member of $f^{-1}\{b_i\}$ $(i = 1, 2)$ then $a_1 + a_2$ is the
least member of $f^{-1}\{b_1 + b_2\}$. Since for any $t \in u/v$, $(t,1) = (t,0) + (v,1)$ it is enough
to show that $f^{-1}\{(v,1)\}$ has a least member. f is surjective, so there exists a $w \in F(X)$
with $f(w) = (v,1)$. Define

$$\overline{w} = w \cdot \prod\{x \in X : (v,1) \le f(x)\} \cdot \prod\{\beta_h(b) : b \in L - I \text{ and } v < b\}.$$

Clearly $f(\overline{w}) = (v,1)$, and if

$$S = \{p \in F(X) : (v,1) \le f(p) \text{ implies } \overline{w} \le p\}$$

then $X \subseteq S$ and S is closed under meets. We need to show that S is also closed under joins. Let $p, q \in S$ and suppose that $(v, 1) \leq f(p + q) = f(p) + f(q)$. Note that by construction of $L[I]$, $r + s \in I \times \{1\}$ implies $r \in I \times \{1\}$ or $s \in I \times \{1\}$, so if $f(p + q) \in I \times \{1\}$, then $(v, 1) \leq f(p)$ or $(v, 1) \leq f(q)$, whence $\overline{w} \leq p$ or $\overline{w} \leq q$, which certainly implies $\overline{w} \leq p + q$. On the other hand, if $f(p + q) \in L - I$, then $\gamma f(p + q) = f(p + q)$, and so $\overline{w} \leq \beta_h \gamma f(p + q) \leq p + q$. Therefore f is lower bounded by $\beta_f : L[I] \hookrightarrow F(X)$, where

$$\beta_f(b) = \begin{cases} \beta_h \gamma(b) & \text{if } b \in (L - I) \cup I \times \{0\} \\ \overline{w} + \beta_h \gamma(b) & \text{if } b \in I \times \{1\}. \end{cases}$$

A dual argument shows that f is also upper bounded, hence $L[I] \in \mathcal{B}_F$. $\square$

Let $W(L)$ be the set of all (W)-failures of the lattice L (see Lemma 2.15) and define

$$\mathcal{I}_W(L) = \{c + d/ab : (a, b, c, d) \in W(L)\}.$$

LEMMA 2.42 (Day [77]). *If L is a lattice that fails (W), then there exists a lattice $\overline{L}$ and a bounded epimorphism $\rho : \overline{L} \twoheadrightarrow L$ satisfying: For any $(a_1, a_2, a_3, a_4) \in W(L)$ and any $x_i \in \rho^{-1}\{a_i\}$ $(i = 1, 2, 3, 4)$ $x_1 x_2 \not\leq x_3 + x_4$.*

PROOF. For each $I \in \mathcal{I}_W(L)$ we construct the lattice $L[I]$ and denote by γ_I the natural epimorphism from $L[I]$ onto L. Note that γ_I is bounded with the upper and lower bounds of $\gamma^{-1}\{b\}$ given by

$$\alpha_I(b) = \begin{cases} b & \text{if } b \in L - I \\ (b, 1) & \text{if } b \in I \end{cases} \qquad \beta_I(b) = \begin{cases} b & \text{if } b \in L - I \\ (b, 0) & \text{if } b \in I \end{cases}$$

respectively. Let L' be the product of all the $L[I]$ as I ranges through $\mathcal{I}_W(L)$, and let $\pi_I : L' \twoheadrightarrow L[I]$ be the Ith projection map. Recall that for $f, g : L' \to L$ we can define a sublattice of L' by

$$\text{Eq}(f, g) = \{x \in L' : f(x) = g(x)\}.$$

Let $\overline{L} = \bigcap\{\text{Eq}(\gamma_I \pi_I, \gamma_J \pi_J) : I, J \in \mathcal{I}_W(L)\}$ and take $\rho : \overline{L} \to L$ to be the restriction of $\gamma_I \pi_I$ to $\overline{L}$. Now, for every $y \in L$, $\gamma_I \alpha_I(y) = y = \gamma_J \alpha_J(y)$, hence the I-tuple $(\alpha_I(y))$ is an element of $\overline{L}$, and clearly $\alpha_\rho(y) = (\alpha_I(y))$ is the greatest element of $\rho^{-1}\{y\}$. Similarly $\beta_\rho(y) = (\beta_I(y))$, and therefore ρ is a bounded epimorphism. To verify the last part of the lemma, it is sufficient to show that for all $(a, b, c, d) \in W(L)$, $\beta_\rho(a)\beta_\rho(b) \not\leq \alpha_\rho(c) + \alpha_\rho(d)$. This is indeed the case, since $c + d/ab = I \in \mathcal{I}_W(L)$ implies

$$\beta_I(a)\beta_I(b) = (ab, 1) \not\leq (c + d, 0) = \alpha_I(c) + \alpha_I(d).$$

$\square$

Note that if $L \in \mathcal{B}_F$, then $\overline{L}$ is a sublattice of a finite product of lattices $L[I]$, hence by Lemmas 2.40 and 2.41, $\overline{L} \in \mathcal{B}_F$.

THEOREM 2.43 (Day [77]). *For any lattice L, there is a lattice $\hat{L}$ satisfying (W) and a bounded epimorphism $\hat{\rho} : \hat{L} \twoheadrightarrow L$.*

PROOF. Let $L_0 = L$ and, for each $n \in \omega$, let $L_{n+1} = \overline{L}_n$ and $\rho_{n+1} : L_{n+1} \twoheadrightarrow L_n$ be given by the preceding lemma. $\hat{L}$ is defined to be the *inverse limit* of the L_n, ρ_n, i.e. $\hat{L}$ is the sublattice of the product $\bigtimes_{n \in \omega} L_n$ defined by

$$x \in \hat{L} \qquad \text{if and only if} \qquad \rho_{n+1}(x_{n+1}) = x_n \quad \text{for all } n \in \omega$$

where $x_i \in L_i$ is the image of x under the projection $\pi_i : \bigtimes_{n \in \omega} L_n \twoheadrightarrow L_i$. We claim that $\hat{L}$ satisfies (W). Suppose $a, b, c, d \in \hat{L}$ with $a, b \not\leq c + d$ and $ab \not\leq c, d$. Then there exist indices j, k, l, m such that

$$a_j b_j \not\leq c_j, \qquad a_k b_k \not\leq d_k, \qquad a_l \not\leq c_l + d_l \quad \text{and} \quad b_m \not\leq c_m + d_m.$$

Since each ρ_i is order-preserving, we have that for any $i \geq \max\{j, k, l, m\}$,

$$a_i, b_i \not\leq c_i + d_i \qquad \text{and} \qquad a_i b_i \not\leq c_i, d_i.$$

Now if $a_i b_i \not\leq c_i + d_i$, then $ab \not\leq c + d$ and we are done. If $a_i b_i \leq c_i + d_i$, then by the previous lemma $a_{i+1} b_{i+1} \not\leq c_{i+1} + d_{i+1}$, and again $ab \not\leq c + d$. Hence $\hat{L}$ satisfies (W).

Let $\hat{\rho} = \pi_0 | \hat{L} : \hat{L} \to L$, let $\overline{\alpha}_0 = \overline{\beta}_0$ be the identity map on $L_0 = L$, and for $n \geq 1$ define the maps $\overline{\alpha}_n, \overline{\beta}_n : L \hookrightarrow L_n$ by

$$\overline{\alpha}_n = \alpha_{\rho_n} \overline{\alpha}_{n-1} \qquad \text{and} \qquad \overline{\beta}_n = \beta_{\rho_n} \overline{\beta}_{n-1}.$$

Then it is easy to check that for $y \in L$ the sequences $(\overline{\alpha}_n(y))$ and $(\overline{\beta}_n(y))$ are the greatest and least elements of $\hat{\rho}^{-1}\{y\}$ respectively, hence $\hat{\rho}$ is a bounded epimorphism. $\square$

THEOREM 2.44 (Day [77]). *$\mathcal{L}$ is generated by the class of all splitting lattices.*

PROOF. Let $L = F_{\mathcal{D}}(3)$, the free distributive lattice on three generators, say x, y, z, and consider the lattice $\hat{L}$ constructed in the preceding theorem. L is a finite distributive lattice, hence $L \in \mathcal{B}_F$, and it follows that $\hat{L} \in (\mathcal{B}_F)^{\mathcal{V}}$. Choose elements $\hat{x}, \hat{y}, \hat{z} \in \hat{L}$ which map to x, y, z under $\hat{\rho} : \hat{L} \twoheadrightarrow L$. Since the set $X = \{x, y, z\}$ satisfies (W2') and (W3'), so does the set $\hat{X} = \{\hat{x}, \hat{y}, \hat{z}\}$. In addition $\hat{L}$ satisfies (W), hence the sublattice of $\hat{L}$ generated by $\hat{X}$ is isomorphic to $F_{\mathcal{L}}(3)$. By a well known result of Whitman [42], the free lattice on countably many generators is a sublattice of $F_{\mathcal{L}}(3)$, and therefore $F_{\mathcal{L}}(\omega) \in (\mathcal{B}_F)^{\mathcal{V}}$. The result now follows. $\square$

The two statements of the following corollary were proven equivalent to the above theorem by A. Kostinsky (see McKenzie [72]).

COROLLARY 2.45

(i) *$F_{\mathcal{L}}(n)$ is weakly atomic for each $n \in \omega$.*

(ii) *For any proper subvariety $\mathcal{V}$ of $\mathcal{L}$, there is a splitting pair $(\mathcal{V}_1, \mathcal{V}_2)$ of $\mathcal{L}$ such that $\mathcal{V} \subseteq \mathcal{V}_1$.*

PROOF. (i) By the above theorem $F_{\mathcal{L}}(n)$ is a subdirect product of splitting lattices S_i ($i \in I$). Let $f : F_{\mathcal{L}}(n) \hookrightarrow \bigtimes_{i \in I} S_i$ be the subdirect representation, and suppose r/s is a nontrivial quotient of $F_{\mathcal{L}}(n)$. Then for some index $i \in I$, $\pi_i f(r) \neq \pi_i f(s)$. Since S_i is finite, we can choose a prime quotient $p/q \subseteq \pi_i f(r)/\pi_i f(s)$. By Theorem 2.25 $\pi_i f : F_{\mathcal{L}}(n) \twoheadrightarrow S_i$

is a bounded epimorphism, so there exists a greatest preimage v of q and a least preimage u of p, and it is easy to check that $u + v/v$ is a prime subquotient of r/s (see proof of Theorem 2.25).

(ii) This is an immediate consequence of the preceding theorem, since $\mathcal{L} = (\mathcal{B}_F)^{\mathcal{V}}$ if and only if every proper subvariety of $\mathcal{L}$ does not contain all splitting lattices. $\square$

Note that if $F_{\mathcal{L}}(n)$ is weakly atomic for $n \in \omega$, then by Corollary 2.26 $F_{\mathcal{L}}(n)$ is a subdirect product of splitting lattices, hence $F_{\mathcal{L}}(n) \in (\mathcal{B}_F)^{\mathcal{V}}$ for each $n \in \omega$. This clearly implies $\mathcal{L} = (\mathcal{B}_F)^{\mathcal{V}}$.

Using some of the results of this section, we prove one last characterization of finite bounded lattices.

THEOREM 2.46 (Day [79]). *A finite lattice L is bounded if and only if there is a sequence of lattices* $\mathbf{1} = L_0, L_1, \ldots, L_{n+1} = L$ *and a sequence of quotients* $u_0/v_0, \ldots, u_n/v_n$ *with* $u_i/v_i \subseteq L_i$ *such that* $L_{i+1} \cong L_i[u_i/v_i]$ ($i = 0, 1, \ldots, n$).

PROOF. The reverse implication follows from Lemma 2.41, since the trivial lattice $\mathbf{1}$ is obviously bounded.

To prove the forward implication, let θ be an atom in $\mathrm{Con}(L)$. We need only show that L can be obtained from $L_n = L/\theta$ by finding a suitable quotient u_n/v_n in L_n such that $L_n[u_n/v_n] \cong L$. Since L_n is again a finite bounded lattice, we can then repeat this process to obtain $L_{n-1}, L_{n-2}, \ldots, L_0 = \mathbf{1}$.

By Theorem 2.39 the map $p \mapsto \mathrm{con}(p, p_*)$ is a bijection from $J(L)$ to $J(\mathrm{Con}(L))$, and since $\theta \in J(\mathrm{Con}(L))$, there exists a unique $p \in J(L)$ with $\theta = \mathrm{con}(p, p_*)$. L is semidistributive, so by the dual of Lemma 2.36 (iii) we can find a unique meet irreducible $m \in L$ such that $m^*/m \searrow p/p_*$, where m^* is the unique cover of m. We claim that

(1) m/p_* transposes bijectively up onto m^*/p, and

(2) $x\theta y$ if and only if $x = y$ or $\{x, y\} = \{z, p + z\}$ for some $z \in m/p_*$.

Letting $u_n = m/\theta$ and $v_n = p/\theta$, we then have $L_n[u_n/v_n] \cong L$.

To prove (1), suppose $x \in m/p_*$ but $x < (p + x)m$. Then we can find $q \in J(L)$ such that $q \le (p + x)m$ and $q \not\le x$. Now $p_*\theta p$ implies $x\,\theta\,(p + x)m$, which in turn implies $q_*\theta q$. Since the map $p \mapsto \mathrm{con}(p, p_*)$ is one-one, this forces $p = q \le m$, a contradiction. Dually one proves that for $x \in m^*/p$, $x = mx + p$.

Since L is a finite lattice we need only check the forward implication of (2) for pairs $(x, y) \in \theta$ of the form $x \prec y$. Clearly $\mathrm{con}(x, y) \le \mathrm{con}(p, p_*)$, and since $\mathrm{con}(p, p_*)$ is an atom of $\mathrm{Con}(L)$, equality holds. This means that p is the unique join irreducible for which $p/p_* \nearrow y/x$, and therefore $\{x, y\} = \{x, p + x\}$. The reverse implication follows from the observation that if $x = z$, say, and $z \in m/p_*$, then $x \ge p_*$, $x \not\ge p$ and $y = p + x$. This implies that $p/p_* \nearrow y/x$, whence $x\theta y$. $\square$

2.5 Finite lattices that satisfy (W)

We conclude this chapter with a result about finite lattices that satisfy Whitman's condition (W), and some remarks about finite sublattices of a free lattice.

THEOREM 2.47 (Davey and Sands[77]). *Suppose f is an epimorphism from a finite lattice K onto a lattice L. If L satisfies* (W), *then there exists an embedding $g : L \hookrightarrow K$ such that fg is the identity map on L.*

PROOF. Let f be the epimorphism from K onto L. Since K is finite, f is bounded, so we obtain the meet preserving map $\alpha_f : L \hookrightarrow K$ and the join preserving map $\beta_f : L \hookrightarrow K$ (see Lemma 2.17(v)). Let M be the collection of all join preserving maps $\gamma : L \to K$ which are pointwise below α_f (i.e. $\gamma(b) \leq \alpha_f(b)$ for all $b \in L$). M is not empty since $\beta_f \in M$. Now define a map $g : L \to K$ by $g(b) = \sum\{\gamma(b) : \gamma \in M\}$. g is clearly join preserving and is in fact the largest element of M (in the pointwise order). Also, since $\beta_f(b) \leq g(b) \leq \alpha_f(b)$ for all $b \in L$, we have

$$b = f\beta_f(b) \leq fg(b) \leq f\alpha_f(b) = b$$

which implies that fg is the identity map on L. It remains to show that g is meet preserving, then g is the desired embedding of L into K.

Suppose $g(ab) \neq g(a)g(b)$ for some $a, b \in L$. Since g is order preserving, we actually have $g(ab) < g(a)g(b)$. Define $h : L \to M$ by

$$h(x) = \begin{cases} g(x) & \text{if } ab \not\leq x \\ g(x) + g(a)g(b) & \text{if } ab \leq x. \end{cases}$$

Then $h \notin M$ because $h(ab) = g(a)g(b) > g(ab)$, but h is pointwise below α_f since for $ab \leq x$ we have $h(x) = g(x) + g(a)g(b) \leq \alpha_f(x) + \alpha_f(a)\alpha_f(b) = \alpha_f(x) + \alpha_f(ab) = \alpha_f(x)$. It follows that h is not join preserving, so there exist $c, d \in L$ such that $h(c + d) \neq h(c) + h(d)$. From the definition of h we see that this is only possible if $ab \leq c + d$, $ab \not\leq c$ and $ab \not\leq d$. Thus (W) implies that $a \leq c + d$ or $b \leq c + d$. However, either one of these conditions leads to a contradiction, since then

$$h(c + d) = g(c + d) + g(a)g(b) = g(c + d) = g(c) + g(d) = h(c) + h(d).$$

$\square$

Actually the result proved in Davey and Sands [77] is somewhat more general, since it suffices to require that every chain of elements in K is finite.

Finite sublattices of a free lattice. Another result worth mentioning is that any finite semidistributive lattice which satisfies Whitman's condition (W) can be embedded in a free lattice. This longstanding conjecture of Jónsson was finally proved by Nation [83]. Following an approach originally suggested by Jónsson, Nation proves that a finite semidistributive lattice L which satisfies (W) cannot contain a cycle. By Corollary 2.35 and Theorem 2.39 L is bounded, and it follows from Theorem 2.19 that L can be embedded in a free lattice. (Note that (W) fails in the lattice of Figure 2.7)

Of course any finite sublattice of a free lattice is semidistributive and satisfies (W) (Corollary 2.16, Lemma 2.30). So in particular Nation's result shows that the finite sublattices of free lattices can be characterized by the first-order conditions (SD^+), $(SD^\cdot)$ and (W).

Chapter 3

Modular Varieties

3.1 Introduction

Modular lattices were studied in general by Dedekind around 1900, and for quite some time they were referred to as Dedekind lattices. The importance of modular lattices stems from the fact that many algebraic structures give rise to such lattices. For example the lattice of normal subgroups of a group and the lattice of subspaces of a vector space and a projective space (projective geometry) are modular. The Jordan Hölder Theorem of group theory depends only on the (semi-) modularity of normal subgroup lattices and the theorem of Kuroš and Ore holds in any modular lattice.

Projective spaces play an important role in the study of modular varieties because their subspace lattices provide us with infinitely many subdirectly irreducible (complemented) modular lattices of arbitrary dimension. They also add a geometric flavor to the study of modular lattices.

The Arguesian identity was introduced by Jónsson [53] (see also Schützenberger [45]). It implies modularity and is a lattice equivalent of Desargues' Law for projective spaces. Some of the results about Arguesian lattices are discussed in Section 3.2, but to keep the length of this presentation within reasonable bounds, most proofs have been omitted.

As we have mentioned before McKenzie [70] and Baker [69] (see also Wille [72] and Lee [85]) showed independently that the lattice Λ of all lattice subvarieties has 2^ω members. Moreover, Baker's proof shows that the lattice $\Lambda_\mathcal{M}$ of all modular lattice subvarieties contains the Boolean algebra 2^ω as a sublattice.

Continuous geometries, as introduced by von Neumann [60], are complemented modular lattices and von Neumann's coordinatization of these structures demonstrates an important connection between rings and modular lattices. Using the notion of an n-frame and its associated coordinate ring (due to von Neumann), Freese [79] shows that the variety $\mathcal{M}$ of all modular lattices is not generated by its finite members. Herrmann [84] extends this result by showing that $\mathcal{M}$ is not even generated by its members of finite length.

The structure of the bottom end of $\Lambda_\mathcal{M}$ is investigated in Grätzer [66] and Jónsson [68], where it is shown that the variety $\mathcal{M}_3$ generated by the diamond M_3 is covered by exactly two varieties, $\mathcal{M}_4$ and $\mathcal{M}_{3^2}$. Furthermore, Jónsson [68] proved that above $\mathcal{M}_4$ we have a chain of varieties $\mathcal{M}_n$, each generated by a finite modular lattice of length 2, such that $\mathcal{M}_{n+1}$ is the only join irreducible cover of $\mathcal{M}_n$. Hong [72] adds further detail to this picture

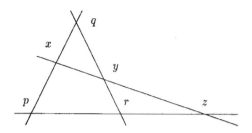

Figure 3.1

by showing, among other things, that $\mathcal{M}_{3^2}$ has exactly five join irreducible covers. The methods developed by Jónsson and Hong have proved to very useful for the investigation of modular varieties generated by lattices of finite length and / or finite width (=maximal number of pairwise incomparable elements). Freese [**72**] extends these methods and gives a complete description of the variety generated by all modular lattices of width 4.

3.2 Projective Spaces and Arguesian Lattices

We begin with a discussion of projective spaces, since many of the results about modular varieties make use of some of the properties of these structures. A some of the results reviewed here will also be used in Chapter 6.

Definition of a projective space. In this section we will be concerned with pairs of sets (P, L), where P is a set of *points* and a collection L of subsets of P, called *lines*. If a point $p \in P$ is an element of a line $l \in L$, then we say that p *lies on* l, and l *passes through* p. A set of points is *collinear* if all the points lie on the same line. A *triangle* is an ordered triple of noncollinear (hence distinct) points (p, q, r).

(P, L) is said to be a *projective space* (sometimes also called projective geometry) if it satisfies:

(P1) each line contains at least two points;

(P2) any two distinct points p and q are contained in *exactly one* line (denoted by $\overline{\{p, q\}}$);

(P3) for any triangle (p, q, r), if a line intersects two of the lines $\overline{\{p, q\}}$, $\overline{\{p, r\}}$ or $\overline{\{q, r\}}$ in distinct points, then it meets the third side (i.e. coplanar lines intersect, see Figure 3.1).

The two simplest projective spaces, which have no lines at all, are $(\emptyset, \emptyset)$ and $(\{p\}, \emptyset)$, while $(\{p, q\}, \{\{p, q\}\})$ and $(\{p, q, r\}, \{\{p, q, r\}\})$ have one line each, and

$$(\{p, q, r\}, \{\{p, q\}, \{p, r\}, \{q, r\}\})$$

has three lines.

The last two examples show that we can have different projective spaces defined on the same set of points. However we will usually be dealing only with one space (P, L) at a time which we then simply denote by the letter P.

A *subspace* S of a projective space P is a subset S of P such that every line which passes through any two distinct points of S is included in S (i.e. $p, q \in S$ implies $\overline{\{p,q\}} \subseteq S$, where we define $\overline{\{p,p\}} = \{p\}$). The collection of all subspaces of P is denoted by $\mathcal{L}(P)$.

Projective spaces and modular geometric lattices. A lattice L is said to be *upper-semimodular* or simply *semimodular* if $a \prec b$ in L implies $a + c \prec b + c$ or $a + c = b + c$ for all $c \in L$. Clearly every modular lattice is semimodular.

A *geometric lattice* is a semimodular algebraic lattice in which the compact elements are exactly the finite joins of atoms. The next theorem summarizes the connection between projective spaces and (modular) geometric lattices.

THEOREM 3.1 *Let P be an arbitrary projective space. Then*

(i) *$(\mathcal{L}(P), \subseteq)$ is a complete modular lattice;*

(ii) *associated with every modular lattice M is a projective space $P(M)$, where $P(M)$ is the set of all atoms of M, and a line through two distinct atoms p and q is the set of atoms below $p + q$ (i.e. $\overline{\{p,q\}} = \{r \in P(M) : r \leq p + q\}$);*

(iii) *$P(\mathcal{L}(P)) \cong P$;*

(iv) *for any modular lattice M, if M' is the sublattice of M generated by the atoms of M, and $\mathcal{I}M'$ is the ideal lattice of M', then $\mathcal{I}M' \cong \mathcal{L}(P(M))$;*

(v) *$\mathcal{L}(P)$ is a modular geometric lattice;*

(vi) *$\mathcal{L}(P(M)) \cong M$ for any modular geometric lattice M*

PROOF. (i) $\mathcal{L}(P)$ is closed under arbitrary intersections and $P \in \mathcal{L}(P)$, hence $\mathcal{L}(P)$, ordered by inclusion, is a complete lattice. For $S, T \in \mathcal{L}(P)$ the join can be described by

$$S + T = \bigcup \{\overline{\{p,q\}} : p \in S, \quad q \in T\}$$

(here we use (P3), see [**GLT**] p.203). Suppose $R \in \mathcal{L}(P)$ and $R \supseteq T$. To prove $\mathcal{L}(P)$ modular, we need only show that $R(S + T) \subseteq RS + T$. Let $r \in R(S + T)$. Then $r \in R$ and $r \in S + T$, which implies $r \in \overline{\{p,q\}}$ for some $p \in S$, $q \in T \subseteq R$. If $r = q$ then $r \in T$, and if $r \neq q$ then $p \in \overline{\{r,q\}} \subseteq R$ (by (P2)), hence $p \in RS$. In either case $r \in RS + T$ as required.

(ii) We have to show that $P(M)$ satisfies (P1), (P2) and (P3). (P1) holds by construction, and (P2) follows from the fact that, by modularity, the join of two atoms covers both atoms. To prove (P3), suppose (p, q, r) form a triangle, and x, y are two distinct points (atoms) such that $x \leq p + q$ and $y \leq q + r$ (see Figure 3.1). It suffices to show that $\overline{\{x,y\}} \cap \overline{\{p,r\}} \neq \emptyset$. Since p, q, r are noncollinear, $p + q + r$ covers $p + r$ by (upper semi-) modularity. Also $x + y \leq p + q + r$, hence $x + y = (x + y)(p + q + r)$, which covers or equals $(x + y)(p + r)$ by (lower semi-) modularity. If $x + y = (x + y)(p + r)$, then $x + y \leq p + r$, and since $x + y$ and $p + r$ are elements of height 2, we must have $x + y = p + r$. In this case $\overline{\{x,y\}} = \overline{\{p,r\}}$ and there intersection is certainly nonempty. If $x + y \succ (x + y)(p + r) = z$, then z must be an atom, and is in fact the point of intersection of $\overline{\{x,y\}}$ and $\overline{\{p,r\}}$.

(iii) A point $p \in P$ corresponds to the one element subspace (atom) $\{p\} \in \mathcal{L}(P)$, and it is easy to check that this map extends to a correspondence between the lines of P and $\mathcal{L}(P)$.

(iv) Each ideal of M' is generated by the set of atoms it contains, every subspace of $P(M)$ is the set of atoms of some ideal of M', and the (infinite) meet operations (intersection) of both lattices are preserved by this correspondence. Hence the result follows.

(v) By (iii) $P \cong P(M)$ for some modular lattice M, hence (iv) implies that $\mathcal{L}(P) \cong \mathcal{I}M'$. It is easy to check that $\mathcal{I}M'$ is a geometric lattice, and modularity follows from (i).

Now (vi) follows from (iv) and the observation that if M is a geometric lattice, then $\mathcal{I}M' \cong M$. $\qquad\square$

LEMMA 3.2 (Birkhoff [35']). *Every geometric lattice is complemented.*

PROOF. Let L be a geometric lattice and consider any nonzero element $x \in L$. By Zorn's Lemma there exists an element $m \in L$ that is maximal with respect to the property $xm = 0_L$. We want to show $x + m = 1_L$. Every element of L is the join of all the atoms below it, so if $x + m < 1_L$, then there is an atom $p \not\leq x + m$, and by semimodularity $m \prec m + p$. We show that $x(m + p) = 0_L$, which then contradicts the maximality of m, and we are done. Suppose $x(m + p) > 0_L$. Then there is an atom $q \leq x(m + p)$, and $q \not\leq m$ since $q \leq x$. Again by semimodularity $m \prec m + q$. Also $q \leq m + p$, hence $m \prec m + q \leq m + p$, and together with $m \prec m + p$ we obtain $m + q = m + p$. But this implies $p \leq m + q \leq x + m$, a contradiction. $\qquad\square$

In fact MacLane [38] showed that every geometric lattice is relatively complemented (see [GLT] p.179).

The next theorem is a significant result that is essentially due to Frink [46], although Jónsson [54] made the observation that the lattice L is in the same variety as K.

THEOREM 3.3 *Let $\mathcal{V}$ be a variety of lattices. Then every complemented modular lattice $K \in \mathcal{V}$ can be 0,1-embedded in some modular geometric lattice $L \in \mathcal{V}$.*

PROOF. Let $M = \mathcal{F}K$ be the filter lattice of K, ordered by reverse inclusion. Then M satisfies all the identities which hold in K, hence M is modular and $M \in \mathcal{V}$. For L we take the subspace lattice of the projective space $P(M)$ associated with M. Note that the points of $P(M)$ are the maximal (proper) filters of K. By Theorem 3.1 (v), L is a modular geometric lattice, and by (iv) $L \cong \mathcal{I}M'$, which implies that L is also in $\mathcal{V}$. Define a map $f : K \to L$ by

$$f(x) = \{F \in P(M) : x \in F\}$$

for each $x \in K$. It is easy to check that $f(x)$ is in fact a subspace of $P(M)$, that $f(0_K) = \emptyset$, $f(1_K) = P(M)$ and that f is meet preserving, hence isotone. To conclude that f is also join preserving, it is therefore sufficient to show that $f(x + y) \subseteq f(x) + f(y)$. This is trivial for x or y equal to 0_K, so suppose $x, y \neq 0_K$ and $F \in f(x + y)$. Then $x + y \in F$, and we have to show that there exist two maximal filters $G \in f(x)$, $H \in f(y)$ such that $F \leq G + H$ (i.e. $F \supseteq G + H$). If $x \in F$, then we simply take $F = G$, and H as any maximal filter containing y, and similarly for $y \in F$ (here, and subsequently, we use Zorn's Lemma to extend any filter to a maximal filter). Thus we may assume that $x, y \notin F$. Further we may assume that $xy = 0_K$, since if $xy > 0_K$, then we let y' be a relative complement of xy in the quotient $y/0_K$ (it is easy to see that every complemented modular lattice is relatively complemented). Clearly $xy' = xyy' = 0_K$, $x + y' = x + xy + y' = x + y$ and any filter that does not contain y must also exclude y', so we can replace y by y'.

Now $y \notin F$ implies $[y) < [y) + F$, where $[y)$ is the principal filter generated by y. Hence by modularity, we see that

$$[0_K) = [x) \cdot [y) < [x) \cdot ([y) + F)$$

(else $[y)$, $[y) + F$ and $[x)$ generate a pentagon). So there is a maximal filter $G \leq [x) \cdot ([y) + F)$, whence it follows that $x \in G$ and $[y) + F = [y) + F + G$. This time $x \notin F$ gives $G < G + F$, and to avoid a pentagon, we must have $[0_K) = [y) \cdot G < [y) \cdot (G + F)$. Hence there is a maximal filter $H \leq [y) \cdot (G + F)$, and $x \in G$, $y \in H$ and $xy = 0_K$ shows that $G \neq H$. Consequently F, G and H are three distinct atoms of $\mathcal{F}K$, and since $H \leq G + F$, they generate a diamond. Thus $F \leq G + H$ as required.

To see that f is one-one, suppose $x \not\leq y$, and let x' be a relative complement of xy in $x/0_K$. If F is a maximal filter containing x', then $F \in f(x)$ but $F \notin f(y)$ since $x'y = 0_K$. Therefore $f(x) \neq f(y)$.   □

The above result is not true if we allow K to be an arbitrary modular lattice. Hall and Dilworth [44] construct a modular lattice that cannot be embedded in any complemented modular lattice.

Coordinatization of projective spaces. The *dimension* of a subspace is defined to be the cardinality of a minimal generating set. This is equal to the height of the subspace in the lattice of all subspaces. If it is finite, then it is one greater than the usual notion of Euclidian dimension, since a line is generated by a minimum of two points. A two-dimensional projective (sub-) space is called a *projective line* and a three-dimensional one is called a *projective plane*.

It is easy to characterize the subspace lattices of projective lines: they are all the (modular) lattices of length 2, excluding the three element chain. Note that except for the four element Boolean algebra, these lattices are all simple. A projective space in which every line has at least three points is termed *nondegenerate*. A simple geometric argument shows that the lines of a nondegenerate projective space all have the same number of points.

Nondegenerate projective spaces are characterized by the fact that their subspace lattices are directly indecomposable (not the direct product of subspace lattices of smaller projective spaces) and, in the light of the following theorem, they form the building blocks of all other projective spaces.

THEOREM 3.4 (Maeda [51]). *Every (modular) geometric lattice is the product of directly indecomposable (modular) geometric lattices.*

A proof of this theorem can be found in [GLT] p.180. There it is also shown that a directly indecomposable modular geometric lattice is subdirectly irreducible (by Lemma 1.13, it will be simple if it is finite dimensional).

An important type of nondegenerate projective space is constructed in the following way:

Let D be a *division ring* (i.e. a ring with unit, in which every nonzero element has a multiplicative inverse), and let V be an α-dimensional vector space over D. (For $\alpha = n$

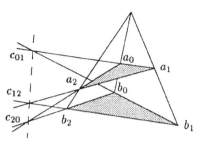

Figure 3.2

finite, V is isomorphic to $_DD^n$, otherwise V is isomorphic to the vector subspace of $_DD^\alpha$ generated by the set $\{e_\gamma : \gamma \in \alpha\}$, where the coordinates of the e_γ are all zero except for a 1 in the γth position.) It is not difficult to check that the lattice $\mathcal{L}(V, D)$ of all vector subspaces of V over D is a modular geometric lattice, so by Theorem 3.1, $\mathcal{L}(V, D)$ determines a projective space P such that $\mathcal{L}(V, D) \cong \mathcal{L}(P)$. Clearly P has dimension α, and the points of P are the one-dimensional vector subspaces of V. Note that P is nondegenerate, for if $p_u = \{au : a \in D\}$ and $p_v = \{av : a \in D\}$ are two distinct points of P (i.e. $u, v \in V, u \neq av$ for any $a \in D$), then the line through these two points must contain the point p_{u-v}, which is different from p_u and p_v (else $u - v = av$, say, giving $u = (a+1)v$ and therefore $p_u = p_v$). Observe also that the number of points on each line (=number of one-dimensional subspaces in any two-dimensional subspace) is $|D| + 1$. The smallest nondegenerate projective space is obtained from $\mathcal{L}(Z_2{}^3, Z_2)$ where Z_2 is the two element field. The subspace lattice, denoted by P_2, is given in Figure 3.7.

We say that a nondegenerate projective space P can be *coordinatized* if $\mathcal{L}(P) \cong \mathcal{L}(V, D)$ for some vector space V over some division ring D. To answer the question which projective spaces can be coordinatized, we need to recall Desargues' Law.

Two triangles $a = (a_0, a_1, a_2)$ and $b = (b_0, b_1, b_2)$ in a projective space P are *centrally perspective* if $\overline{\{a_i, a_j\}} \neq \overline{\{b_i, b_j\}}$ and for some point p the points a_i, b_i, p are collinear $(i, j \in \{0, 1, 2\})$. If we think of the points a_i, b_i as atoms of the lattice $\mathcal{L}(P)$, then we can express this condition by

$$(a_0 + b_0)(a_1 + b_1) \le a_2 + b_2.$$

The triangles are said to be *axially perspective* if the points c_0, c_1, c_2 are collinear, where $c_k = (a_i + a_j)(b_i + b_j)$, $\{i, j, k\} = \{0, 1, 2\}$ (see Figure 3.2). This can be expressed by

$$c_2 \le c_0 + c_1.$$

Desargues' Law states that if two triangles are centrally perspective then they are also axially perspective. A projective space which satisfies Desargues' Law is said to be *Desarguesian*.

It is a standard result of projective geometry that every projective space associated with a vector subspace lattice is Desarguesian.

Conversely, we have the classical coordinatization theorem of projective geometry, due to Veblen and Young [10] for the finite dimensional case and Frink [46] in general.

THEOREM 3.5 *Let P be a nondegenerate Desarguesian projective space of dimension $\alpha \geq 3$. Then there exists a division ring D, unique up to isomorphism, such that $\mathcal{L}(P) \cong \mathcal{L}(D^\alpha, D)$.*

For a proof of this theorem and further details, the reader should consult [ATL] p.111 or [GLT] p.208. Here we remark only that to construct the division ring D which coordinatizes P we may choose an arbitrary line l of P and define D on the set $l - \{p\}$ where p is any point of l. The 0 and 1 of D may also be chosen arbitrarily, and the addition and multiplication are then defined with reference to the lattice operations in $\mathcal{L}(P)$. This leads to the following observation:

LEMMA 3.6 *Let P and Q be two nondegenerate Desarguesian projective spaces of dimension ≥ 3 and let D_P and D_Q be the corresponding division rings which coordinate them. If $\mathcal{L}(P)$ can be embedded in $\mathcal{L}(Q)$ such that the atoms of $\mathcal{L}(P)$ are mapped to atoms of $\mathcal{L}(Q)$, then D_P can be embedded in D_Q.*

It is interesting to note that projective spaces of dimension 4 or more automatically satisfy Desargues' Law ([GLT] p.207), hence any noncoordinatizable projective space is either degenerate, or a projective plane that does not satisfy Desargues' Law, or a projective line that has $k + 1$ points, where k is a finite number that is not a prime power.

Arguesian lattices. The lattice theoretic version of Desargues' Law can be generalized to any lattice L by considering arbitrary triples $a, b \in L^3$ (also referred to as triangles in L) instead of just triples of atoms. We now show that under the assumption of modularity this form of Desargues' Law is equivalent to the *Arguesian* identity:

$$(a_0 + b_0)(a_1 + b_1)(a_2 + b_2) \leq a_0(a_1 + d) + b_0(b_1 + d)$$

where d is used as an abbreviation for

$$d = c_2(c_0 + c_1) = (a_0 + a_1)(b_0 + b_1)((a_1 + a_2)(b_1 + b_2) + (a_0 + a_2)(b_0 + b_2)).$$

A lattice is said to be *Arguesian* if it satisfies this identity.

LEMMA 3.7 *Let $p = (a_0 + b_0)(a_1 + b_1)(a_2 + b_2)$, then*

(i) *the identity $p \leq a_0(a_1 + d) + b_0$ is equivalent to the Arguesian identity,*

(ii) *every Arguesian lattice is modular and*

(iii) *to check whether the Arguesian identity holds in a modular lattice, it is enough to consider triangles $a' = (a'_0, b'_0, c'_0)$ and $b' = (b'_0, b'_1, b'_2)$ which satisfy*

$$(*) \qquad a'_i + b'_i = a'_i + p' = b'_i + p', \quad (i = 0, 1, 2)$$

where p' is defined in the same manner as p.

PROOF. Since we always have $b_0(b_1 + d) \leq b_0$, the Arguesian identity clearly implies $p \leq a_0(a_1+d)+b_0$. Conversely, let L be a lattice which satisfies the identity $p \leq a_0(a_1+d)+b_0$. We first show that L is modular. Given $u, v, w \in L$ with $u \leq w$, let $a_0 = v$, $b_0 = u$ and

$a_1 = a_2 = b_1 = b_2 = w$. Then $p = (v + u)w$ and $d = w$, whence the identity implies $(v + u)w \leq vw + u$. Since $u + vw \leq (u + v)w$ holds in any lattice, we have equality, and so L is modular. This proves (ii).

To complete (i), observe that p and d are unchanged if we swop the a_i's with their corresponding b_i's, hence we also have $p \leq b_0(b_1+d)+a_0$. Combining these two inequalities gives

$$p \leq (a_0(a_1 + d) + b_0)(b_0(b_1 + d) + a_0)$$
$$= a_0(a_1 + d) + b_0(b_0(b_1 + d) + a_0)$$
$$= a_0(a_1 + d) + a_0b_0 + b_0(b_1 + d)$$

by modularity. Also $a_0b_0 \leq c_2$, c_1 shows $a_0b_0 \leq d$ and therefore $a_0b_0 \leq a_0(a_1 + d)$. This means we can delete the term a_0b_0 and obtain the Arguesian identity.

Now let $a, b \in L^3$ and define $a_i' = a_i(b_i + p)$, $b_i' = b_i(a_i + p)$. Since we are assuming that L is modular,

$$a_i' + b_i' = a_i(b_i + p) + b_i(a_i + p) = (a_i(b_i + p) + b_i)(a_i + p)$$
$$= (b_i + p)(a_i + b_i)(a_i + p)$$
$$= (b_i + p)(a_i + p) = (b_i + p)a_i + p = a_i' + p$$
$$= b_i(a_i + p) + p = b_i' + p.$$

Thus $p' = (a_0' + p)(a_1' + p)(a_2' + p) \geq p$, while $a_i' \leq a_i$ and $b_i' \leq b_i$ imply $p' \leq p$. So we have $p = p'$, and condition $(*)$ is satisfied. If the Arguesian identity holds for a', b' and we define d' in the same way as d, then clearly $d' \leq d$ and

$$p = p' \leq a_0'(a_1' + d') + b_0' \leq a_0(a_1 + d) + b_0,$$

hence the identity holds for the triangles a, b. □

THEOREM 3.8 *If a modular lattice L satisfies Desargues' Law then L is Arguesian. Conversely, if L is Arguesian, then L satisfies Desargues' Law.*

PROOF. Let $a_0, a_1, a_2, b_0, b_1, b_2 \in L$, $p = (a_0 + b_0)(a_1 + b_1)(a_2 + b_2)$, $c_k = (a_i + a_j)(b_i + b_j)$, $(\{i, j, k\} = \{0, 1, 2\})$ and $d = c_2(c_0 + c_1)$ as before. By part (iii) of the preceding lemma we may assume that

$$(*) \qquad a_i + b_i = a_i + p = b_i + p \qquad i = 0, 1, 2.$$

Define $\bar{b}_2 = b_2 + b_0(a_1 + b_1)$. The following calculation shows that the triangles (a_0, a_1, a_2) and $(b_0, b_1, \bar{b}_2)$ are centrally perspective:

$$(a_0 + b_0)(a_1 + b_1) = (p + b_0)(a_1 + b_1) \qquad \text{by } (*)$$
$$= p + b_0(a_1 + b_1) \qquad \text{by modularity}$$
$$\leq a_2 + b_2 + b_0(a_1 + b_1) = a_2 + \bar{b}_2.$$

Therefore Desargues' Law implies that

$$c_2 \leq (a_1 + a_2)(b_1 + \bar{b}_2) + (a_2 + a_0)(\bar{b}_2 + b_0)$$
$$= (a_1 + a_2)(b_1 + b_2 + b_0(a_1 + b_1)) + (a_2 + a_0)(b_2 + b_0)$$
$$= (a_1 + a_2)(b_2 + (b_1 + b_0)(a_1 + b_1)) + c_1$$
$$= (a_1 + a_2)(b_1 + b_2 + a_1(b_0 + b_1)) + c_1$$
$$= (a_1 + a_2)(b_1 + b_2) + a_1(b_0 + b_1) + c_1 = c_0 + c_1 + a_1(b_0 + b_1),$$

whence $c_2 = c_2(c_0 + c_1 + a_1(b_0 + b_1)) = c_2(c_0 + c_1) + a_1(b_0 + b_1) = d + a_1(b_0 + b_1)$. It follows that

$$
\begin{aligned}
a_1 + d = a_1 + c_2 &= a_1 + (a_0 + a_1)(b_0 + b_1) \\
&= (a_1 + b_1 + b_0)(a_0 + a_1) \\
&= (a_1 + p + b_0)(a_0 + a_1) \qquad \text{by } (*) \\
&= (a_1 + p + a_0)(a_0 + a_1) \qquad \text{by } (*) \\
&= (a_0 + a_1) \geq a_0,
\end{aligned}
$$

so we finally obtain $a_0(a_1 + d) + b_0 = a_0 + b_0 \geq p$.

Conversely, suppose L is Arguesian (hence modular) and (a_0, a_1, a_2), (b_0, b_1, b_2) are centrally perspective, i.e. $(a_0 + b_0)(a_1 + b_1) \leq a_2 + b_2$. Let $\bar{c}_0 = (a_1 + a_2)(c_1 + c_2)$ and take $a_0' = \bar{c}_0$, $a_1' = b_1$, $a_2' = a_1$, $b_0' = c_1$, $b_1' = b_0$, $b_2' = a_0$ in the (equivalent form of the) Arguesian identity $p' \leq a_0'(a_1' + d') + b_0'$. We claim that under these assignments $p' = c_2$ and $a_0'(a_1' + d') + b_0' \leq c_0 + c_1$ from which it follows that the two triangles are axially perspective. Firstly,

$$
\begin{aligned}
\bar{c}_0 + c_1 &= (a_1 + a_2 + c_1)(c_1 + c_2) \\
&= (a_1 + (a_0 + a_2)(b_0 + a_2 + b_2))(c_1 + c_2) \\
&\geq (a_1 + (a_0 + a_2)(b_0 + (a_0 + b_0)(a_1 + b_1)))(c_1 + c_2) \\
&= (a_1 + (a_0 + a_2)(a_0 + b_0)(b_0 + a_1 + b_1))(c_1 + c_2) \\
&\geq (a_1 + a_0(b_0 + a_1 + b_1))(c_1 + c_2) \\
&= (a_1 + a_0)(b_0 + a_1 + b_1)(c_1 + c_2) \\
&\geq (a_0 + a_1)(b_0 + b_1)(c_1 + c_2) = c_2
\end{aligned}
$$

so $p' = (\bar{c}_0 + c_1)(b_1 + b_0)(a_1 + a_0) = c_2$. Secondly,

$$
\begin{aligned}
d' &= (\bar{c}_0 + b_1)(c_1 + b_0)((\bar{c}_0 + a_1)(c_1 + a_0) + (b_1 + a_1)(b_0 + a_0)) \\
&\leq (b_0 + b_2)((a_0 + a_2)(a_1 + a_2) + (a_0 + b_0)(a_1 + b_1)) \\
&\leq (b_0 + b_2)((a_0 + a_2)(a_1 + a_2) + b_2) \qquad \text{(by central persp.)} \\
&= (b_0 + b_2)(a_0 + a_2)(a_1 + a_2) + b_2 = c_1(a_1 + a_2) + b_2
\end{aligned}
$$

which implies

$$
\begin{aligned}
a_0'(a_1' + d') + b_0' &= \bar{c}_0(b_1 + d') + c_1 \\
&\leq (a_1 + a_2)(b_1 + b_2 + c_1(a_1 + a_2)) + c_1 \\
&= (a_1 + a_2)(b_1 + b_2) + c_1(a_1 + a_2) + c_1 = c_0 + c_1
\end{aligned}
$$

□

The first statement of this theorem appeared in Grätzer, Jónsson and Lakser [73], and the converse is due to Jónsson and Monk [69]. In [GLT] p.205 it is shown that for any Desarguesian projective plane P the atoms of $\mathcal{L}(P)$ satisfy the Arguesian identity and that this implies that $\mathcal{L}(P)$ is Arguesian. Hence it follows from the preceding theorem that P is Desarguesian if and only if $\mathcal{L}(P)$ satisfies (the generalized version of) Desargues' Law.

Since modularity is characterized by the exclusion of the pentagon N, which is isomorphic to its dual, it follows that the class of all modular lattices $\mathcal{M}$ is *self-dual* (i.e. $M \in \mathcal{M}$ implies that the dual of M is also in $\mathcal{M}$). The preceding theorem can be used to prove the corresponding result for the variety of all Arguesian lattices.

LEMMA 3.9 (Jónsson[72]). *The variety of all Arguesian lattices is self-dual.*

PROOF. For modular lattices the Arguesian identity is equivalent to Desargues' Law by Theorem 3.8. Let L be an Arguesian lattice and denote its dual by $\overline{L}$. Lemma 3.7 (ii) implies that L is modular, and by the above remark, so is $\overline{L}$. We show that the dual of Desargues' Law holds in L, i.e. for all $x_0, x_1, x_2, y_0, y_1, y_2 \in L$

$$(*) \qquad x_0 y_0 + x_1 y_1 \geq x_2 y_2$$

implies that

$$(**) \qquad x_0 x_1 + y_0 y_1 \geq (x_1 x_2 + y_1 y_2)(x_0 x_2 + y_0 y_2).$$

Then $\overline{L}$ satisfies Desargues' Law and is therefore Arguesian.

Assume $(*)$ holds, and let $a_0 = x_0 x_2$, $a_1 = y_0 y_2$, $a_2 = x_0 y_0$, $b_0 = x_1 x_2$, $b_1 = y_1 y_2$, $b_2 = x_1 y_1$ and $c_k = (a_i + a_j)(b_i + b_j)$ ($\{i, j, k\} = \{0, 1, 2\}$). Then

$$(a_0 + b_0)(a_1 + b_1) = (x_0 x_2 + x_1 x_2)(y_0 y_2 + y_1 y_2) \leq x_2 y_2 \leq a_2 + b_2$$

by $(*)$, so it follows from Desargues' Law that $c_2 \leq c_0 + c_1$. But $c_0 \leq y_0 y_1$, $c_1 \leq x_0 x_1$ and c_2 equals the right hand side of $(**)$. Therefore $(**)$ is satisfied. $\qquad\square$

So far we have only considered the most basic properties of Arguesian lattices. Extensive research has been done on these lattices, and many important results have been obtained in recent years. We mention some of the results now.

Recall that the collection of all equivalence relations (partitions) on a fixed set form an algebraic lattice, with intersection as meet. If two equivalence relations permute with each other under the operation of composition then their join is simply the composite relation. A lattice is said to be *linear* if it can be embedded in a lattice of equivalence relations in such a way that any pair of elements is mapped to a pair of permuting equivalence relations. (These lattices are also referred to as lattices that have a *type 1 representation*, see [GLT] p.198). An example of a linear lattice is the lattice of all normal subgroups of a group (since groups have permutable congruences), and similar considerations apply to the "subobject" lattices associated with rings, modules and vectorspaces.

Jónsson [53] showed that any linear lattice is Arguesian, and posed the problem whether the converse also holds. A recent example of Haiman [86] shows that this is not the case, i.e. there exist Arguesian lattices which are not linear.

Most of the modular lattices which have been studied are actually Arguesian. The question how a modular lattice fails to be Arguesian is investigated in Day and Jónsson [89].

Pickering [84] [a] proves that there is a non-Arguesian, modular variety of lattices, all of whose members of finite length are Arguesian. This result shows that Arguesian lattices cannot be characterized by the exclusion of a finite list of lattices or even infinitely many lattices of finite length. For reasons of space the details of these results are not included here.

The cardinality of $\Lambda_{\mathcal{M}}$. In this section we discuss the result of Baker [69] which shows that there are uncountably many modular varieties. We begin with a simple observation about finite dimensional modular lattices.

LEMMA 3.10 *Let L and M be two modular lattices, both of dimension $n < \omega$. If a map $f : L \hookrightarrow M$ is one-one and order-preserving then f is an embedding.*

PROOF. We have to show that

$$(*) \qquad f(x+y) = f(x) + f(y) \qquad \text{and} \qquad f(xy) = f(x)f(y).$$

Since f is assumed to be order-preserving, $f(x+y) \geq f(x) + f(y)$, $f(xy) \leq f(x)f(y)$ and equality holds if x is comparable with y. For x, y noncomparable we use induction on the length of the quotient $x + y/xy$.

Observe firstly that if $u \succ v$ in L then u, v are successive elements in some maximal chain of L, and since M has the same dimension as L and f is one-one it follows that $f(u) \succ f(v)$. If the length of $x + y/xy$ is 2, then x and y cover xy and x, y are both covered by $x + y$, so $(*)$ holds in this case. Now suppose the length of $x + y/xy$ is $n > 2$ and $(*)$ holds for all quotients of length $< n$. Then either $x + y/x$ or x/xy has length ≥ 2. By modularity $x/xy \cong x + y/y$, and by symmetry we can assume that there exists x' such that $x < x' < x + y$. The quotients $x' + y/x'y$ and $x + x'y/x(x'y) = x'/xy$ have length $< n$, hence

$$f(x') = f(x + x'y) = f(x) + f(x')f(y) = (f(x) + f(y))f(x').$$

It follows that $f(x') \leq f(x) + f(y)$ and so $f(x+y) = f(x'+y) = f(x') + f(y) \leq f(x) + f(y)$. Similarly $f(xy) \geq f(x)f(y)$.   □

Let P be a finite partially ordered set and define $\mathbf{N}(P)$ to be the class of all lattices that do not contain a subset order-isomorphic to P. For example if $\mathbf{5}$ is the linearly ordered set $\{0, 1, 2, 3, 4\}$ then $\mathbf{N}(\mathbf{5})$ is the class of all lattices of length ≤ 4.

LEMMA 3.11 *For any finite partially ordered set P*

  (i) $\mathbf{N}(P)$ *is closed under ultraproducts, sublattices and homomorphic images;*

  (ii) *any subdirectly irreducible lattice in the variety $\mathbf{N}(P)^{\mathcal{V}}$ is a member of $\mathbf{N}(P)$.*

PROOF. (i) The property of not containing a finite partially ordered set can be expressed as a first-order sentence and is therefore preserved under ultraproducts. If L is a lattice and a sublattice of L contains a copy of P, then of course so does L. Finally, if a homomorphic image of L contains P then for each minimal $p \in P$ choose an inverse image $\overline{p} \in L$, and thereafter choose an inverse image $\overline{q}$ of each $q \in P$ covering a minimal element in P such that $\overline{q} \geq \sum \{\overline{p} : p \leq q, \ p \in P\}$. Proceeding in this way one obtains a copy of P in L.

  (ii) This is an immediate consequence of Corollary 1.5.   □

THEOREM 3.12 (Baker [69]). *There are uncountably many modular lattice varieties.*

PROOF. Let Π be the set of all prime numbers, and for each $p \in \Pi$ denote by F_p the p-element Galois field. Let $L_p = \mathcal{L}(F_p^3, F_p)$ and observe that each L_p is a finite subdirectly irreducible lattice since it is the subspace lattice of a finite nondegenerate projective space. We also let $\mathcal{A}$ be the class of all Arguesian lattices of length ≤ 4.

Now define a map f from the set of all subsets of Π to $\Lambda_{\mathcal{M}}$ by

$$f(S) = \mathcal{A}^{\mathcal{V}} \cap \bigcap \{\mathbf{N}(L_q)^{\mathcal{V}} : q \notin S\}.$$

We claim that f is one-one. Suppose $S, T \subseteq \Pi$ and $p \in S - T$. Then $f(T) \subseteq \mathbf{N}(L_p)^{\mathcal{V}}$ and since $L_p \notin \mathbf{N}(L_p)$ and L_p is subdirectly irreducible it follows from the preceding lemma that $L_p \notin f(T)$.

On the other hand we must have $L_p \in f(S)$ since $L_p \notin \mathbf{N}(L_q)$ for some $q \notin S$ would imply that L_p contains a subset order-isomorphic to L_q. By Lemma 3.10 L_q is actually a sublattice of L_p and it follows from Lemma 3.6 that F_q is a subfield of F_p. This however is impossible since $q \neq p \in S$. $\square$

By a more detailed argument one can show that the map f above is in fact a lattice embedding, from which it follows that $\Lambda_{\mathcal{M}}$ contains a copy of $\mathbf{2}^\omega$ as a sublattice.

3.3 n-Frames and Freese's Theorem

Products of projective modular lattices. By a projective modular lattice we mean a lattice which is projective in the variety of all modular lattices.

LEMMA 3.13 (Freese[**76**]). *If A and B are projective modular lattices with greatest and least element then $A \times B$ is a projective modular lattice.*

PROOF. Let f be a homomorphism from a free modular lattice F onto $A \times B$, and choose elements $u, v \in F$ such that $f(u) = (1_A, 0_B)$ and $f(v) = (0_A, 1_B)$. By Lemma 2.9 it suffices to produce an embedding $g : A \times B \hookrightarrow F$ such that fg is the identity on $A \times B$.

Clearly f followed by the projection π_A onto the first coordinate maps the quotient u/uv onto A. Assuming that A is projective modular, there exists an embedding $g_A : A \hookrightarrow u/uv$ such that $\pi_A f g_A$ is the identity on A. Similarly, if B is projective modular, there exists an embedding $g_B : B \hookrightarrow v/uv$ such that $\pi_B f g_B = \mathrm{id}_B$. Define g by $g(a, b) = g_A(a) + g_B(b)$ for all $(a, b) \in A \times B$. Then g is join preserving, and clearly fg is the identity on $A \times B$. To see that g is also meet preserving, observe that by the modularity of F

$$g(a, b) = g_A(a) + vu + g_B(b) = (g_A(a) + v)u + g_B(b) = (g_A(a) + v)(u + g_B(b)).$$

Hence

$$g(a, b)g(c, d) = (g_A(a) + v)(u + g_B(b))(g_A(c) + v)(u + g_B(d))$$
$$= (g_A(a)g_A(c) + v)(u + g_B(b)g_B(d)) = g(ac, bd),$$

where the middle equality follows from the fact that in a modular lattice the map $t \mapsto t + v$ is an isomorphism from u/uv to $u + v/v$. $\square$

Von Neumann n-frames. Let $\{a_i : i = 1, \ldots, n\}$ and $\{c_{1j} : j = 2, \ldots, n\}$ be subsets of a modular lattice L for some finite $n \geq 2$. We say that $\phi = (a_i, c_{1j})$ is an *n-frame* in L if the sublattice of L generated by the a_i is a Boolean algebra $\mathbf{2}^n$ with atoms $a_1, \ldots, a_n$, and for each $j = 2, \ldots, n$ the elements a_1, c_{1j}, a_j generate a diamond in L (i.e. $a_1 + c_{1j} = a_j + c_{1j} = a_1 + a_j$ and $a_1 c_{1j} = a_j c_{1j} = a_1 a_j$). The top and bottom element of the Boolean algebra are denoted by 0_ϕ ($= a_1 a_2$) and 1_ϕ ($= \sum_{i=1}^n a_i$) respectively, but they need not equal the top and bottom of L (denoted by 0_L and 1_L). If they do, then ϕ is called a *spanning n-frame*.

If the elements $a_1, \ldots, a_n \in L$ are the atoms of a sublattice isomorphic to $\mathbf{2}^n$, then they are said to be *independent* over $0 = a_1 a_2$. If L is modular this is equivalent to the conditions $a_i \neq 0$ and $a_i \sum_{j \neq i} a_j = 0$ for all $i = 1, \ldots, n$ (see [**GLT**] p.167).

The index 1 in c_{1j} indicates that an n-frame determines further elements c_{ij} for distinct $i, j \neq 1$ as follows: let $c_{j1} = c_{1j}$ and

$$c_{ij} = (a_i + a_j)(c_{i1} + c_{1j}).$$

These elements fit nicely into the n-frame, as is shown by the next lemma.

LEMMA 3.14 *Let $\phi = (a_i, c_{1j})$ be an n-frame in a modular lattice and suppose c_{ij} is defined as above. Then, for distinct $i, j \in \{1, \ldots, n\}$*

(i) $a_i + c_{ij} = a_i + a_j = c_{ij} + a_j$;

(ii) $c_{ij} \sum_{r \neq j} a_r = 0_\phi$;

(iii) $a_i \sum_{r \neq k} c_{kr} = 0_\phi$ *for any fixed index k;*

(iv) a_i, c_{ij}, a_j *generate a diamond;*

(v) $c_{ij} = (a_i + a_j)(c_{ik} + c_{kj})$ *for any k distinct from i, j.*

PROOF. (i) Using modularity and the n-frame relations, we compute

$$
\begin{aligned}
a_i + c_{ij} &= a_i + (a_i + a_j)(c_{i1} + c_{1j}) \\
&= (a_i + a_j)(a_i + c_{i1} + c_{1j}) \\
&= (a_i + a_j)(a_i + a_1 + a_j) = a_i + a_j.
\end{aligned}
$$

The second part follows by symmetry.

(ii) We first show that $c_{ij} \sum_{r \neq j} a_r \leq a_i$.

$$
\begin{aligned}
a_i + c_{ij} \sum_{r \neq j} a_r &= (a_i + c_{ij}) \sum_{r \neq j} a_r \quad \text{by modularity since } i \neq j \\
&= (a_i + a_j) \sum_{r \neq j} a_r = a_i \quad \text{since the } a_i\text{'s generate } 2^n.
\end{aligned}
$$

Hence if $i = 1$ then $0_\phi \leq c_{1j} \sum_{r \neq j} a_r \leq c_{1j} a_1 = 0_\phi$. The general case will follow in the same way once we have proved (iv).

(iii) We first fix $i = k = 1$ and show that $a_1 \sum_{r=2}^m c_{1r} \leq \sum_{r=2}^{m-1} c_{1r}$ for $3 \leq m \leq n$.

$$
\begin{aligned}
a_1 \sum_{r=2}^m c_{1r} + \sum_{r=2}^{m_1} c_{1r} &= (c_{12} + \ldots + c_{1m})(a_1 + c_{12} + \ldots + c_{1m-1}) \\
&= (c_{12} + \ldots + c_{1m})(a_1 + a_2 + \ldots + a_{m-1}) \\
&= c_{12} + \ldots + c_{1m-1} + c_{1m} \sum_{r=1}^{m-1} a_r \\
&= c_{12} + \ldots + c_{1m-1} + 0_\phi \quad \text{by part (ii).}
\end{aligned}
$$

Thus $0_\phi \leq a_1 \sum_{r=2}^n c_{1r} \leq a_1 \sum_{r=2}^{n-1} c_{1r} \leq \ldots \leq a_1 c_{12} = 0_\phi$.

Let $e = \sum_{r=2}^n c_{1r}$ and suppose $i \neq 1$. Then $c_{1i} + a_i e = (c_{1i} + a_i)e = (c_{1i} + a_1)e = c_{1i} + a_1 e = c_{1i} + 0_\phi$, so $a_i e \leq a_i c_{1i} = 0_\phi$. Hence (iii) holds for $k = 1$ and any i.

Now (iv) follows from (i) and the calculation

$$
0_\phi \leq a_i c_{ij} = a_i(a_i + a_j)(c_{i1} + c_{ij}) = a_i(c_{1i} + c_{1j}) \leq a_i e = 0_\phi.
$$

Therefore (ii) holds in general. Using this one can show in the same way as for $k = 1$, that $a_k \sum_{r=k+1}^m c_{kr} \leq \sum_{r=k+1}^{m-1} c_{kr}$ for $k + 1 \leq m \leq n$ and, letting $c' = \sum_{r=k+1}^n c_{kr}$, $a_k(c' + \sum_{r=1}^m c_{kr}) \leq c' + \sum_{r=1}^{m-1} c_{rk}$ for $1 \leq m \leq k - 1$. Assuming $i \neq k$, let $e = \sum_{r \neq k} c_{rk}$. Then one shows as before that $c_{ki} + a_i e = c_{ki}$, whence $a_i e = 0_\phi$. Thus (iii) holds in general.

For $k = 1$ (v) holds by definition. Suppose $i = 1 \neq j, k$.

$$
\begin{aligned}
(a_1 + a_j)(c_{1k} + c_{kj}) &= (a_1 + a_j)(c_{1k} + (a_k + a_j)(c_{k1} + c_{1j})) \\
&= (a_1 + a_j)(c_{1k} + a_k + a_j)(c_{k1} + c_{1j}) \\
&= (a_1 + a_j)(a_1 + a_k + a_j)c_{k1} + c_{1j} = 0_\phi + c_{1j} \quad \text{by (ii).}
\end{aligned}
$$

The case $j = 1 \neq i, k$ is handled similarly. Finally suppose that $i, j, k, 1$ are all distinct and note that $c_{ik} \leq a_i + a_j + a_k$.

$$
\begin{aligned}
(a_i + a_j)(c_{ik} + c_{kj}) &= (a_i + a_j)((a_i + a_j + a_k)(c_{i1} + c_{1k}) + c_{kj}) \\
&= (a_i + a_j)(c_{i1} + c_{1k} + c_{kj}) \\
&= (a_i + a_j)(c_{i1} + (a_1 + a_j)(c_{1k} + c_{kj})) \\
&= (a_i + a_j)(c_{i1} + c_{1j}) = c_{ij}
\end{aligned}
$$

$\square$

A concept equivalent to that of an n-frame is the following: A modular lattice L contains an n-*diamond* $\delta = (a_1, \ldots, a_n, e)$ if the a_i are independent over $0_\delta = a_1 a_2$ and e is a relative complement of each a_i in $1_\delta/0_\delta$ ($1_\delta = \sum_{j=1}^{n} a_j$). The concept of an n-diamond is due to Huhn [72] (he referred to it as an $(n-1)$-diamond).

Note that although e seems to be a special element relative to the a_i, this is not really true since any n elements of the set $\{a_1, \ldots, a_n, e\}$ are independent, and the remaining element is a relative complement of all the others.

LEMMA 3.15 Let $\delta = (a_i, e)$ be an n-diamond and define $c_{1j} = e(a_1 + a_j)$, then $\phi_\delta = (a_i, c_{1j})$ is an n-frame. Conversely, if $\phi = (a_i, c_{1j})$ is an n-frame and $e = \sum_{j=2}^{n} c_{1j}$ then $\delta_\phi = (a_i, e)$ is an n-diamond. Furthermore $\phi_{\delta_\phi} = \phi$ and $\delta_{\phi_\delta} = \delta$.

PROOF. Since e is a relative complement for each a_i in $1_\delta/0_\delta$, $a_1 e(a_1 + a_j) = 0_\delta = a_j e(a_1 + a_j)$ and

$$
a_1 + e(a_1 + a_j) = (a_1 + e)(a_1 + a_j) = 1(a_1 + a_j) = a_1 + a_j = a_j + e(a_1 + a_j),
$$

so $\phi_\delta = (a_i, e(a_1 + a_j))$ is an n-frame.

Conversely, if $e = \sum_{j=2}^{n} c_{1j}$ then $a_i e = 0_\phi$ by Lemma 3.14 (iii) and

$$
\begin{aligned}
a_i + e &= c_{12} + \ldots + a_i + c_{1i} + \ldots + c_{1n} \\
&= c_{12} + \ldots + a_i + a_1 + \ldots + c_{1n} \\
&= a_1 + \ldots + a_n = 1.
\end{aligned}
$$

Hence $\delta_\phi = (a_i, e)$ is an n-diamond.

Also $e(a_1 + a_j) = c_{1j} + (\sum_{r \neq j} c_{1r})(a_1 + a_j) = c_{1j}$ since $(\sum_{r \neq j} c_{1r})(a_1 + a_j) = 0_\phi$ can be proved similar to Lemma 3.14 (iii). Finally, if $\delta = (a_i, e)$ is any n-diamond, and we let $e' = \sum_{j=2}^{n} c_{1j}$, then $e' \leq e$ and in fact $e' = e' + (a_1 + a_2)e = (e' + a_1 + a_2)e = 1_\delta e = e$. $\square$

LEMMA 3.16 (Freese[76]). Suppose $\beta = (a_i, e)$ is an $(n+1)$-tuple of elements of a modular lattice such that the a_i form an independent set over $0_\beta = a_1 a_2$, $0_\beta \leq e \leq 1_\beta = join_{i=1}^{n} a_1$ and e is incomparable with each a_i. Define (i ranges over $1, \ldots, n$)

$$
\begin{aligned}
b = \sum a_i e &\leq e \leq c = \textstyle\prod(a_i + e) \\
d = \sum(a_i + b)c &= b + \sum a_i c = \sum a_i c \qquad \text{and} \\
\beta^* = (a_i + b, e), \quad \beta_* &= (a_i c, ed), \quad \beta^*_* = ((a_i + b)c, ed).
\end{aligned}
$$

(i) If $a_i + e = 1$ for all i then β^* is an n-diamond in $1/b$.

(ii) If $a_i e = 0_\beta \neq a_i c$ for all i then β_* is an n-diamond in $d/0_\beta$.

(iii) *If* $b \neq (a_i + b)c$ *for all* i *then* $\beta^*{}_*$ *is an* n-*diamond in* d/b.

PROOF. (i) Since $b \leq e$ and $a_i \not\leq e$ we have $a_i + b \neq b$ for all i. The following calculation shows that the $a_i + b$ are independent over b:

$$(a_i + b)\sum_{j \neq i}(a_j + b) = b + a_i(\sum_{j \neq i} a_j + \sum_{k=1}^n a_k e)$$
$$= b + a_i(\sum_{j \neq i} a_j + a_i e)$$
$$= b + a_i \sum_{j \neq i} a_j + a_i e$$
$$= b + 0_\beta + a_i e = b.$$

Furthermore $(a_i + b) + e = a_i + b = 1$ by assumption, and $(a_i + b)e = b + a_i e = b$.

(ii) Since $a_i c \neq 0_\beta$ and $0_\beta \leq a_i c \sum_{j \neq i} a_j c \leq a_i \sum_{j \neq i} a_j = 0_\beta$, the $a_i c$ are independent over 0_β. Also $e \leq c$ and $a_i e \leq a_i c \leq d$ imply $(a_i c)(ed) = a_i ed = a_i e = 0_\beta$ by assumption, and $a_i c + ed = (a_i c + e)d = c(a_i + e)d = cd = d$.

Now (iii) follows from (i) and (ii). $\qquad\qquad\qquad\qquad\qquad\qquad\qquad\qquad\qquad\qquad\quad\Box$

Suppose M and L are two modular lattices and f is a homomorphism from M to L. If $\phi = (a_i, c_{1j})$ is an n-frame in M and the elements $f(a_i)$, $f(c_{1j})$ are all distinct, then $(f(a_i), f(c_{1j}))$ is an n-frame in L (since the diamonds generated by a_1, c_{1i}, a_i are simple lattices). Risking a slight abuse of notation, we will denote this n-frame by $f(\phi)$. Of course similar considerations apply to n-diamonds.

The next result shows that n-diamonds (and hence n-frames) can be "pulled back" along epimorphisms.

COROLLARY 3.17 (Huhn [72], Freese [76]). *Let* M *and* L *be modular lattices and let* $f : M \twoheadrightarrow L$ *be an epimorphism. If* $\delta = (a_i, e)$ *is an* n-*diamond in* L *then there is an* n-*diamond* $\hat{\delta} = (\hat{a}_i, \hat{e})$ *in* M *such that* $f(\hat{\delta}) = \delta$.

PROOF. It follows from Lemma 3.13 that 2^n is a projective modular lattice, so we can find $\overline{a}_1, \ldots, \overline{a}_n \in M$ such that $f(\overline{a}_i) = a_i$ and the $\overline{a}_i$ are independent over $\overline{a}_1 \overline{a}_2$. Choose $\overline{e} \in f^{-1}\{e\}$ such that $\overline{a}_1 \overline{a}_2 \leq \overline{e} \leq \sum_{i=1}^n \overline{a}_i$ and let $\beta = (\overline{a}_i, \overline{e})$. Since δ is an n-diamond, each $\overline{a}_i$ is incomparable with $\overline{e}$. Defining $\overline{b}, \overline{c}, \overline{d}$ in the same way as b, c, d in the previous lemma, we see that $f(\overline{b}) = 0_\delta$, $f(\overline{c}) = 1_\delta$ and $f((\overline{a}_i + \overline{b})\overline{c}) = a_i$. Therefore $\overline{b} \neq (\overline{a}_i + \overline{b})\overline{c}$, whence $\hat{\delta} = \beta^*{}_*$ is the required n-diamond. $\qquad\qquad\qquad\qquad\qquad\qquad\Box$

LEMMA 3.18 (Herrmann and Huhn[76]). *Let* $\phi = (a_i, c_{1j})$ *be an* n-*frame in a modular lattice* L *and let* $u_1 \in L$ *satisfy* $0_\phi \leq u_1 \leq a_1$. *Define* $u_i = a_i(u_1 + c_{1i})$ *for* $i \neq 1$ *and* $u = \sum_{i=1}^n u_i$. *Then* $\phi^u = (u + a_i, u + c_{1j})$ *and* $\phi_u = (ua_i, uc_{1j})$ *are* n-*frames in* 1_ϕ *and* $u/0_\phi$ *respectively.*

A proof of this result can be found in Freese [79]. We think of ϕ^u (ϕ_u) as being obtained from ϕ by a reduction over (under) u.

The canonical n-frame. The following example shows that n-frames occur naturally in the study of R-modules:

Let $(R, +, -, \cdot, 0_R, 1_R)$ be a ring with unit, and let $\mathcal{L}(R^n, R)$ be the lattice of all (left-) submodules of the (left-) R-module R^n. We denote the canonical basis of R^n by $e_1, \ldots, e_n$ (i.e. $e_i = (0_R, \ldots, 0_R, 1_R, 0_R, \ldots, 0_R)$ with the 1_R in the ith position), and let

$$a_i = Re_i = \{re_i : r \in R\}$$
$$c_{ij} = R(e_i - e_j) \qquad i, j = 1, \ldots, n \quad i \neq j.$$

Then it is not difficult to check that $\mathcal{L}(R^n, R)$ is a modular lattice and that $\phi_R = (a_i, c_{1j})$ is a (spanning) n-frame in $\mathcal{L}(R^n, R)$, referred to as the *canonical n-frame* of $\mathcal{L}(R^n, R)$.

Definition of the auxillary ring. Let L be a modular lattice containing an n-frame $\phi = (a_i, c_{1j})$ for some $n \geq 3$. We define an *auxillary ring* R_ϕ associated with the frame ϕ as follows:

$$R_\phi = R_{12} = \{x \in L : xa_2 = a_1 a_2 \text{ and } x + a_2 = a_1 + a_2\}$$

and for some $k \in \{3, \ldots, n\}$, $x, y \in R_\phi$

$$\pi(x) = (x + c_{1k})(a_2 + a_k), \qquad \pi'(x) = (x + c_{2k})(a_1 + a_k)$$
$$x \oplus y = (a_1 + a_2)[(x + a_k)(c_{1k} + a_2) + \pi(y)]$$
$$x \ominus y = (a_1 + a_2)[a_k + (c_{2k} + x)(a_2 + \pi'(y))]$$
$$x \odot y = (a_1 + a_2)[\pi(x) + \pi'(y)]$$
$$0_R = a_1, \qquad 1_R = c_{12}.$$

THEOREM 3.19 *If $n \geq 4$, or L is an Arguesian lattice and $n = 3$, then $(R_\phi, \oplus, \ominus, \odot, 0_R, 1_R)$ is a ring with unit, and the operations are independent of the choice of k.*

This theorem is due to von Neumann [60] for $n \geq 4$ and Day and Pickering [83] for $n = 3$. The presentation here is derived from Herrmann [84], where the theorem is stated without proof in a similar form. The proof is long, as many properties have to be checked, and will be omitted here as well. The theorem however is fundamental to the study of modular lattices.

It is interesting to compare the definition of R with the definition D in the classical coordinatization theorem for projective spaces ([**GLT**] p.209). The element a_2 corresponds to the point at infinity, and the operations of addition and multiplication are defined in the same way.

There is nothing special about the indices 1 and 2 in the definition of $R_\phi = R_{12}$. We can replace them throughout by distinct indices i and j to obtain isomorphic rings R_{ij}. For example the isomorphism between R_{12} and R_{1j} ($j \neq 1, 2$) is induced by the projectivity

$$R_{12} \subseteq a_1 + a_2/0 \nearrow a_1 + a_2 + c_{2j}/c_{2j} \searrow a_1 + a_j/0 \supseteq R_{1j}.$$

(Since in a modular lattice every transposition is bijective, it only remains to show that this induced map preserves the respective operations. For readers more familiar with von Neumann's L-numbers, we note that they are $n(n-1)$-tuples of elements $\beta_{ij} \in R_{ij}$, which correspond to each other under the above isomorphisms.)

Coordinatization of complemented modular lattices. The auxillary ring construction is actually part of the von Neumann coordinatization theorem, which we will not use, but mention here briefly (for more detail, the reader is referred to von Neumann [60]).

THEOREM 3.20 *Let L be a complemented modular lattice containing a spanning n-frame $(4 \leq n \in \omega)$ and let R be the auxillary ring. Then L is isomorphic to the lattice $\mathcal{L}_f(R^n, R)$ of all finitely generated submodules of the R-module R^n.*

Notice that if R happens to be a division ring D, then D^n will be a vector space over D, and $\mathcal{L}_f(D^n, D) \cong \mathcal{L}(D^n, D)$. Hence the above theorem extends the coordinatization of (finite dimensional) projective spaces to arbitrary complemented modular lattices containing a spanning n-frame ($n \geq 4$). Moreover, Jónsson [59] [60'] showed that if L is a complemented Arguesian lattice, then the above theorem also holds for $n = 3$. Further generalizations to wider classes of modular lattices appear in Baer [52], Inaba [48], Jónsson and Monk [69], Day and Pickering [83] and Herrmann [84].

Characteristic of an n-frame. Recall that the *characteristic* of a ring R with unit 1_R is the least number $r = \operatorname{char} R$ such that adding 1_R to itself r times equals 0_R. If no such r exists, then $\operatorname{char} R = 0$.

We define a related concept for n-frames as follows:

Let $\phi = (a_i, c_{1j})$ be an n-frame in some modular lattice L ($n \geq 3$), and choose $k \in \{3, \ldots, n\}$. The projectivity

$$a_1 + a_2/0 \nearrow a_1 + a_2 + a_k/a_k \searrow c_{1k} + a_2/0 \nearrow a_1 + a_2 + a_k/c_{2k} \searrow a_1 + a_2/0$$

induces an automorphism α_ϕ on the quotient $a_1 + a_2/0$ given by

$$\alpha_\phi(x) = ((x + a_k)(c_{1k} + a_2) + c_{2k})(a_1 + a_2).$$

Let r be a natural number and denote by α_ϕ^r the automorphism α_ϕ iterated r times. We say that ϕ is an *n-frame of characteristic r* if $\alpha_\phi^r(a_1) = a_1$.

LEMMA 3.21 *Suppose $\phi = (a_i, c_{1j})$ is an n-frame of characteristic r, and R is the auxiliary ring of ϕ. Then the characteristic of R divides r.*

PROOF. By definition $0_R = a_1$ and $1_R = c_{12}$. From the definition of $x \oplus y$ we see that for $x \in R$, $\alpha_\phi(x) = x \oplus 1_R$ (since $\pi(1_R) = (c_{21} + c_{1k})(a_2 + a_k) = c_{2k}$). This also shows that $\alpha_\phi | R$ is independent of the choice of k. The condition $\alpha_\phi^r(a_1) = a_1$ therefore implies

$$0_R \oplus \overbrace{1_R \oplus 1_R \oplus \ldots \oplus 1_R}^{r \text{ terms}} = 0_R$$

whence the result follows.   □

The next result shows that the automorphism α_ϕ is compatible with the operation of reducing an n-frame. For a proof the reader is referred to the original paper.

LEMMA 3.22 (Freese [79]). *Let ϕ, u, ϕ^u and ϕ_u be as in Lemma 3.18. If $x \in a_1 + a_2/0_\phi$ then*

(i) $x + u \in a_1 + a_2 + u/u$ *and* $\alpha_{\phi^u}(x + u) = \alpha_\phi(x) + u$;

(ii) $xu \in a_1 u + a_2 u/0_\phi$ *and* $\alpha_{\phi_u}(xu) = \alpha_\phi(x)u$.

It follows that if ϕ is an n-frame of characteristic r, then so are ϕ^u and ϕ_u. The lemma below shows how one can obtain an n-frame of any given characteristic.

LEMMA 3.23 (Freese [79]). *Let $\phi = (a_i, c_{1j})$ be an n-frame and r any natural number. If we define $a = \alpha_\phi^r(a_1)$, $u_2 = a_2(a + a_1)$, $u_1 = a_1(u_2 + c_{12})$, $u_i = a_i(u_1 + c_{1i})$ and $u = \sum_{i=1}^n u_i$ then ϕ^u is an n-frame of characteristic r.*

PROOF. Note that u_2, defined as above, agrees with the definition of u_2 in Lemma 3.18 since

$$a_2(u_1 + c_{12}) = a_2(a_1(u_2 + c_{12}) + c_{12})$$
$$= a_2(u_1 + c_{12})(u_2 + c_{12})$$
$$= a_2(u_2 + c_{12}) = a_2 c_{12} + u_2 = u_2.$$

Let R be the auxillary ring of ϕ. By definition the elements of R are all the relative complements of a_2 in $a_1 + a_2/0_\phi$. Since $x \in R$ implies $\alpha_\phi(x) = x \oplus 1_R \in R$, it follows that $\alpha_\phi(x)$ is again a relative complement of a_2 in $a_1 + a_2/0_\phi$ (this can also be verified easily from the definition of α_ϕ). Thus $a = \alpha_\phi^r(a_1) \in R$ and $a + a_2 = a_1 + a_2$. By the preceding lemma

$$\alpha_{\phi^u}^r(a_1 + u) = \alpha_\phi^r(a_1) + u$$
$$= a + u + a_2(a + a_1)$$
$$= (a + a_2)(a + a_1) + u$$
$$= (a_1 + a_2)(a + a_1) + u = a + a_1 + u.$$

Also $a_1 + u_2 = (a_1 + a_2)(a_1 + a) = a_1 + a$ shows that $a_1 + u \geq a$, whence $\alpha_{\phi^u}^r(a_1 + u) = a_1 + u$. □

We can now prove the result corresponding to Theorem 3.17 for n-frames of a given characteristic.

THEOREM 3.24 (Freese [79]). *Let M and L be modular lattices and let $f : M \twoheadrightarrow L$ be an epimorphism. If $\phi = (a_i, c_{1j})$ is an n-frame of characteristic r in L, then there is an n-frame $\hat{\phi} = (\hat{a}_i, \hat{c}_{1j})$ of characteristic r in M such that $f(\hat{\phi}) = \phi$.*

PROOF. From Theorem 3.17 we obtain an n-frame $\overline{\phi} = (\overline{a}_i, \overline{c}_{1j})$ in M such that $f(\overline{\phi}) = \phi$. If we let $u_2 = \overline{a}_2(\alpha_{\overline{\phi}}^r(\overline{a}_1) + \overline{a}_1)$ and u be as in the preceding lemma, then we see that $\overline{\phi}^u = (a_i + u, c_{1j} + u)$ is an n-frame of characteristic r in M. Since ϕ has characteristic r by assumption,

$$f(u_2) = f(\overline{a}_2(\alpha_{\overline{\phi}}^r(a_1) + a_1)) = a_2(\alpha_\phi^r(a_1) + a_1) = a_2(a_1 + a_1) = 0_\phi,$$

from which it easily follows that $f(u_1) = f(\overline{a}_1(u_2 + \overline{c}_{12}) = 0_\phi$ and $f(u_i) = 0_\phi$. Therefore $f(u) = 0_\phi$ and $f(\overline{\phi}^u) = \phi$, so we can take $\hat{\phi} = \overline{\phi}^u$. □

$\mathcal{M}$ **is not generated by its finite members.** This is the main result of Freese [79], and follows immediately from the theorem below, where $\mathcal{M}_F$ is the class of all finite modular lattices.

THEOREM 3.25 (Freese [79]). *There exists a modular lattice L such that $L \notin (\mathcal{M}_F)^\mathcal{V}$.*

PROOF. The lattice L is constructed (using a technique due to Hall and Dilworth [44]) as follows:

Let F and K be two countably infinite fields of characteristic p and q respectively, where p and q are distinct primes. Let $L_p = \mathcal{L}(F^4, F)$ be the subspace lattice of the 4-dimensional vector space F^4 over F and let $\phi = (a_i, c_{1j})$ be the canonical 4-frame in L_p (the index $i = 1, 2, 3, 4$ and $j = 2, 3, 4$ through out). Note that ϕ is a spanning 4-frame of characteristic p. Similarly let $L_q = \mathcal{L}(K^4, K)$ with canonical 4-frame $\phi' = (a_i', c_{1j}')$ of characteristic q. Since $|K| = \omega$, there are precisely ω one-dimensional subspaces in the quotient $a_1' + a_2'/0_{\phi'}$, hence $a_1' + a_2'/0_{\phi'} \cong M_\omega$ (the countable two-dimensional lattice). The quotient $1_\phi/a_3 + a_4$ of L_p is isomorphic to $a_1 + a_2/0_\phi$ via the map $x \mapsto x(a_1 + a_2)$ and since

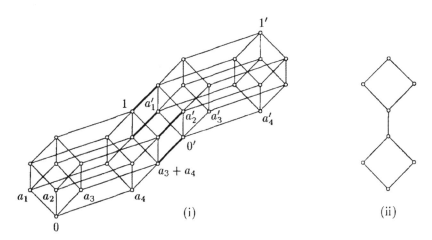

Figure 3.3

$|F| = \omega$, we see that $1_\phi / a_3 + a_4$ is also isomorphic to M_ω. Let $\sigma : 1_\phi / a_3 + a_4 \to a_1' + a_2' / 0_{\phi'}$ be any isomorphism which satisfies

$$\sigma(a_1 + a_3 + a_4) = a_1'$$
$$\sigma(a_2 + a_3 + a_4) = a_2'$$
$$\sigma(c_{12} + a_3 + a_4) = c_{12}'$$

The lattice L is constructed by "loosely gluing" the lattice L_q over L_p via the isomorphism σ, i.e. let L be the disjoint union of L_p and L_q and define $x \le y$ in L if and only if

$$x, y \in L_p \quad \text{and} \quad x \le y \quad \text{in } L_p \text{ or}$$
$$x, y \in L_q \quad \text{and} \quad x \le y \quad \text{in } L_q \text{ or}$$
$$x \in L_p, \; y \in L_q \quad \text{and} \quad x \le z, \; \sigma(z) \le y \quad \text{for some } z \in 1_\phi / a_3 + a_4.$$

Then it is easy to check that L is a modular lattice. (The conditions on σ are needed to make the two 4-frames fit together nicely.) Let D be the finite distributive sublattice of L generated by the set $\{a_i, a_i' : i = 1, 2, 3, 4\}$ (see Figure 3.3 (i)). Notice that D is the product of the four element Boolean algebra and the lattice in Figure 3.3 (ii). Both these lattices are finite projective modular lattices, so by Lemma 3.13 D is a projective modular lattice.

Suppose now that $L \in (\mathcal{M}_F)^\vee = \mathbf{HSP}\mathcal{M}_F$. Then L is a homomorphic image of some lattice $\overline{L} \in \mathbf{SP}\mathcal{M}_F$, and hence $\overline{L}$ is *residually finite* (i.e. a subdirect product of finite lattices). But hereafter we show that any lattice which has L as a homomorphic image cannot be residually finite, and this contradiction will conclude the proof.

Let f be the homomorphism from $\overline{L} \twoheadrightarrow L$. Since D is a projective modular sublattice of L, we can find elements $\overline{a}_i, \overline{a}_i'$ in $\overline{L}$ which generate a sublattice isomorphic to D, and $f(\overline{a}_i) = a_i$, $f(\overline{a}_i') = a_i'$. Let us assume for the moment that

$(*)$ there exist further elements $\overline{c}_{1j}$ and $\overline{c}_{1j}'$ in $\overline{L}$ such that $\overline{\phi} = (\overline{a}_i, \overline{c}_{1j})$ is a 4-frame of characteristic p, $\overline{\phi}' = (\overline{a}_i', \overline{c}_{1j}')$ is a 4-frame of characteristic q and $f(\overline{\phi}) = \phi$, $f(\overline{\phi}') = \phi'$.

If $\overline{L}$ is residually finite, then we can find a finite modular lattice M and a homomorphism $g : \overline{L} \to M$ which maps the (finite) 4-frames $\overline{\phi}$ and $\overline{\phi}'$ in a one to one fashion into M, where we denote them by $\hat{\phi} = (\hat{a}_i, \hat{c}_{1j})$ and $\hat{\phi}'(\hat{a}_i', \hat{c}_{1j}')$ respectively. By Lemmas 3.19 and 3.21 they give rise to two auxiliary rings $R \subseteq \hat{a}_1 + \hat{a}_2/0_{\hat{\phi}}$ and $R' \subseteq \hat{a}_1' + \hat{a}_2'/0_{\hat{\phi}'}$ of characteristic p and q respectively. Since M is a finite lattice, R and R' are finite rings, so $|R| = p^m$ and $|R'| = q^n$ for some $n, m \in \omega$. Also, since $0_R = \hat{a}_1 \neq \hat{c}_{12} = 1_R$, R has at least two elements.

Now in M the elements $\hat{a}_i, \hat{a}_i'$ generate a sublattice $\hat{D} \cong D$, hence $\hat{a}_1 + \hat{a}_2/0_{\hat{\phi}} \nearrow \hat{a}_1' + \hat{a}_2'/0_{\hat{\phi}'}$, $\hat{a}_1 + 0_{\hat{\phi}'} = \hat{a}_1'$ and $\hat{a}_2 + 0_{\hat{\phi}'} = \hat{a}_2'$ (see Figure 3.3 (i)). It follows that the two quotients are isomorphic, and checking the definition of R_ϕ above Theorem 3.19, we see that this isomorphism restricts to an isomorphism between R and R'. Thus $|R| = |R'|$, which is a contradiction, as p and q are distinct primes and $|R| \geq 2$. Consequently $\overline{L}$ is not residually finite, which implies that L is not a member of $(\mathcal{M}_F)^{\mathcal{V}}$.

We now complete the proof with a justification of $(*)$. This is done by adjusting the elements $\overline{a}_i, \overline{a}_i'$ in several steps, thereby constructing the required 4-frames. Since we will be working primarily with elements of $\overline{L}$, we first of all change the notation, denoting the 4-frames ϕ, ϕ' in L by $\hat{\phi} = (\hat{a}_i, \hat{c}_{1j})$, $\hat{\phi}' = (\hat{a}_i', \hat{c}_{1j}')$ and the $\overline{a}_i, \overline{a}_i'$ in $\overline{L}$ by a_i, a_i'. Also the condition that the elements a_i, a_i' generate a sublattice isomorphic to D (Figure 3.3 (i)) will be abbreviated by $D(a_i, a_i')$. To check that $D(a_i, a_i')$ holds, one has to verify that the a_i are independent over $a_1 a_2$, the a_i' are independent over $a_1' a_2' = 0'$, $1/a_3 + a_4 \nearrow a_1' + a_2'/0'$, $a_1' = a_1 + 0'$ and $a_2' = a_2 + 0'$. Actually, once the transposition has been established, it is enough to show that $a_1 \leq a_1'$ and $a_2 \leq a_2'$ since then $0' \leq a_1'(a_2 + 0') \leq a_1' a_2' = 0'$ implies

$$a_1' = (1 + 0')a_1' = (a_1 + a_2 + 0')a_1' = a_1 + (a_2 + 0')a_1' = a_1 + 0',$$

and $a_2' = a_2 + 0'$ follows similarly.

Step 1: Let $1' = \sum_{i=1}^4 a_i'$ and $\hat{e}' = \hat{c}_{12}' + \hat{c}_{13}' + \hat{c}_{14}'$. By Lemma 3.15 $(\hat{a}_i', \hat{e}')$ is a 4-diamond in L. Since $\hat{e}' \in 1_{\hat{\phi}'}/0_{\hat{\phi}'}$, we can choose $e' \in 1'/0'$ such that $f(e') = \hat{e}'$. Clearly e' is incomparable with each a_i'. Defining $b' = \sum a_i' e'$, $c' = \prod(a_i' + e')$, $d' = \sum(a_i' + b')c'$ $(i = 1, 2, 3, 4)$ corresponding to b, c, d in Lemma 3.16 it is easy to check that $b' \leq d' \leq c'$, $(a_i' + b')c' = a_i' c' + b'$, $f(b') = 0_{\hat{\phi}'}$, $f(c') = 1_{\hat{\phi}'} = f(d')$ and $f((a_i' + b')c') = \hat{a}_i'$, so combining part (iii) of that lemma with Lemma 3.15 shows that

$$\overline{\phi}' = (\overline{a}_i', \overline{c}_{1j}') = (a_i' c' + b', (a_1' c' + a_j' c' + b')e' d')$$

is a 4-frame in d'/b' and $f(\overline{\phi}') = \hat{\phi}'$.

Now let $\overline{0} = (a_1 + a_2)b'$ and consider the elements $\overline{a}_i = a_i d' + \overline{0}$. Since $D(\hat{a}_i, \hat{a}_i')$ holds, $\hat{a}_1 + \hat{a}_2 \not\leq 0_{\hat{\phi}'}$, whence $f(\overline{0}) = 0_{\hat{\phi}}$ and $f(\overline{a}_i) = \hat{a}_i$. In particular it follows that $\hat{a}_i \neq \overline{0}$. We show that the $\overline{a}_i$ are independent over $\overline{0}$. Observe that $a_3 + a_4 \leq 0' \leq b' \leq d'$ implies $\overline{a}_3 = a_3 + \overline{0}$ and $\overline{a}_4 = a_4 + \overline{0}$. Since $a_1 \leq a_1'$, $a_2 \leq a_2'$, $d' \leq c'$ and $\overline{\phi}'$ is a 4-frame,

$$\overline{a}_1 \sum_{i \neq 1} \overline{a}_i = (a_1 d' + \overline{0})(a_2 d' + a_3 + a_4 + \overline{0})$$
$$\leq (a_1' c' + b')(a_2' c' + b') = b'.$$

Since $\bar{a}_1 \leq a_1 + a_2$, the left hand side is $\leq 0'$, and the opposite inequality is obvious. Similarly $\bar{a}_2 \sum_{i \neq 2} \bar{a}_i = \bar{0}$. Also

$$\bar{a}_3 \sum_{i \neq 3} \bar{a}_i = (a_3 + \bar{0})(a_1 d' + a_2 d' + a_4 + \bar{0})$$
$$= \bar{0} + a_3(a_1 d' + a_2 d' + a_4 + \bar{0}) = \bar{0}$$

because $a_1 d' + a_2 d' + a_4 + \bar{0} \leq a_1 + a_2 + a_4$, and likewise for $\bar{a}_4 \sum_{i \neq 4} \bar{a}_i = \bar{0}$.

Let $\bar{I} = \sum_{i=1}^4 \bar{a}_i$ and observe that $0_{\bar{\phi}'} = b'$. We proceed to show that $D(\bar{a}_i, \bar{a}'_i)$ holds. Note firstly that $a'_i d' = a'_i c'$ since $a'_i d' = a'_i \sum_{k=1}^4 a'_k c' = a'_i (a'_i c' + \sum_{k \neq i} a'_k c') = a'_i c' + 0'$. Now

$$\bar{I} + b' = a_1 d' + a_2 d' + a_3 + a_4 + \bar{0} + b'$$
$$\leq a'_1 d' + a'_2 d' + b' = a'_1 c' + a'_2 c' + b' = \bar{a}'_1 + \bar{a}'_2,$$

and $a_1 d' + b' \geq a_1 d' + \bar{0} = (a_1 + 0')d' = a'_1 d'$ together with a similar computation for a_2 shows that $\bar{I} + b' = \bar{a}'_1 + \bar{a}'_2$. Furthermore

$$\bar{I} b' = (a_1 d' + a_2 d' + a_3 + a_4 + \bar{0})b'$$
$$= a = 3 + a = 4 + \bar{0} + (a_1 d' + a_2 d')b'$$
$$= a_3 + a_4 + \bar{0} = \bar{a}_3 + \bar{a}_4.$$

Since $\bar{a}_i = a_i d' + (a_1 + a_2)b' \leq a'_i c' + b' = \bar{a}'_i$ we have $D(\bar{a}_i, \bar{a}'_i)$.

Step 2: Using Lemmas 3.18 and 3.23 we now construct a new 4-frame

$$\phi' = (a'_i, c'_{1j}) = \overline{\phi'}^u = (\bar{a}'_i + u, \bar{c}'_{1j} + u),$$

where u is derived from $u_2 = \bar{a}'_2(\alpha^q_{\bar{\phi}}(\bar{a}'_1) + \bar{a}'_1)$. By Lemma 3.23 ϕ' is a 4-frame of characteristic q. Since $\hat{\phi}'$ in L has characteristic q, it follows that $f(u) = 0_{\hat{\phi}}$ and $f(\phi') = \hat{\phi}'$ (see proof of Theorem 3.24). Moreover, if we define $0 = (\bar{a}_1 + \bar{a}_2)u$ and $a_i = \bar{a}_i + 0$ then $f(a_i) = \hat{a}_i$ and calculations similar to the ones in Step 1 show that the a_i, a'_i generate a copy of D.

Step 3: In this step we first construct a new 4-frame $\bar{\phi} = (\bar{a}_i, \bar{c}_{1j})$ derived from the elements a_i of Step 2 such that $f(\bar{\phi}) = \hat{\phi}$. Then we adjust ϕ' accordingly to obtain $\bar{\phi}' = (\bar{a}'_i, \bar{c}'_{1j})$ satisfying $f(\bar{\phi}') = \hat{\phi}'$, $D(\bar{a}_i, \bar{a}'_i)$ and $\bar{c}'_{12} = \bar{c}_{12} + 0_{\bar{\phi}'}$.

Since $\hat{c}_{23} + \hat{c}_{24} \leq \hat{a}_2 + \hat{a}_3 + \hat{a}_4$, it is possible to choose $\bar{e} \in \bar{L}$ such that $f(\bar{e}) = \hat{c}_{23} + \hat{c}_{24}$ and $a_1 a_2 \leq \bar{e} \leq a_2 + a_3 + a_4$. Let $c_{12} = c'_{12}(a_1 + a_2)$ and observe that $f(c_{12}) = \hat{c}'_{12}(\hat{a}_1 + \hat{a}_2) = \hat{c}_{12}$ by the choice of σ in the construction of L. Let $e = c_{12} + \bar{e}$ and define b, c, d as in Lemma 3.16. Since $f(e) = \hat{c}_{12} + \hat{c}_{23} + \hat{c}_{24} = \hat{e}$ is a relative complement of each $\hat{a}_i$, considerations similar to the ones in Step 1 show that

$$\bar{\phi} = (\bar{a}_i, \bar{c}_{1j}) = (a_i c + b, (a_1 c + a_j c + b)ed)$$

is a 4-frame in d/b and $f(\bar{\phi}) = \hat{\phi}$.

$$\text{Let} \quad u'_1 = a_1 c + 0_{\phi'}, \quad u'_i = a'_i(u'_1 + c'_{1i}) \quad u' = \sum_{i=1}^4 u'_i$$
$$\text{and} \quad v'_1 = a_1 e + 0_{\phi'}, \quad v'_i = a'_i(v'_1 + c'_{1i}) \quad v' = \sum_{i=1}^4 v'_i.$$

Then $0_{\phi'} \leq v'_1 \leq u'_1 \leq a_1$ and two applications of Lemma 3.18 show that

$$\bar{\phi}' = (\bar{a}'_i, \bar{c}'_{1j}) = \phi'^{v'}_{u'} = (a'_i u' + v', c'_{1j} u' + v')$$

is a 4-frame in u'/v'. Also since ϕ' has characteristic q, Lemma 3.22 implies that $\overline{\phi}'$ has characteristic q. Furthermore, it is easy to check that $f(u') = 1_{\hat{\phi}'}$ and $f(v') = 0_{\hat{\phi}'}$, whence $f(\overline{\phi}') = \hat{\phi}'$.

We now show that the elements $\overline{a}_i, \overline{a}'_i$ generate a copy of D. From $D(a_i, a'_i)$ we deduce that

$$c_{12} + 0_{\phi'} = c'_{12}(a_1 + a_2) + 0_{\phi'} = c'_{12}(a_1 + a_2 + 0_{\phi'}) = c'_{12}(a'_1 + a'_2) = c'_{12}$$
$$a_1 + c_{12} = a_1 + c'_{12}(a_1 + a_2) = (a_1 + c'_{12})(a_1 + a_2)$$
$$= (a_1 + 0_{\phi'} + c'_{12})(a_1 + a_2) = (a'_1 + a'_2)(a_1 + a_2) = a_1 + a_2.$$

Similarly $a_2 + c_{12} = a_1 + a_2$ and $a_1 c_{12} = a_1 a_2 = a_2 c_{12}$. Since $c_{12} \le e \le c$,

$$c_{12} + a_1 e = (c_{12} + a_1)e = (c_{12} + a_2)e = c_{12} + a_2 e$$
$$c_{12} + a_1 c = c_{12} + a_2 c.$$

Moreover

$$u'_2 = a'_2(a_1 c + 0_{\phi'} + c'_{12}) = a'_2(a_1 c + 0_{\phi'} + c_{12})$$
$$= a'_2(a_2 c + 0_{\phi'} + c_{12}) = a'_2(a_2 c + c'_{12})$$
$$= a_2 c + a'_2 c'_{12} = a_2 c + 0_{\phi'}.$$

A similar calculation yields $v'_2 = a_2 e + 0_{\phi'}$. Now $a_2 c + u'_3 + u'_4 \le a'_2 + a'_3 + a'_4$ implies

$$a'_1 u' = a'_1(a_1 c + a_2 c + 0_{\phi'} + u'_3 + u'_4) = a_1 c + 0_{\phi'},$$

and similarly $a'_2 u' = a_2 c + 0_{\phi'}$. Together with $a_3 c + a_4 c \le v'$ we compute

$$d + v' = \sum_{i=1}^4 a_i c + v' = \sum_{i=1}^4 a_i c + 0_{\phi'} + v'$$
$$= a'_1 u' + a'_2 u' + v' = \overline{a}'_1 + \overline{a}'_2,$$
$$d\, v' = d(a'_1 + a'_2)(a_1 e + a_2 e + 0_{\phi'} + v'_3 + v'_4)$$
$$= d(a_1 e + a_2 e + 0_{\phi'} + (a'_1 + a'_2)(v'_3 + v'_4))$$
$$= d(a_1 e + a_2 e + 0_{\phi'})$$
$$= (a_1 c + a_2 c + a_3 c + a_4 c)0_{\phi'} + a_1 e + a_2 e$$
$$= a_3 c + a_4 c + (a_1 c + a_2 c)0_{\phi'} + a_1 e + a_2 e$$
$$= a_3 c + a_4 c + b = \overline{a}_3 + \overline{a}_4.$$

Since $\overline{a}_1 + v' = a_1 c + b + v' = a_1 c + 0_{\phi'} + v' = a_1 u' + v' = \overline{a}'_1$ and similarly $\overline{a}_2 + v' = \overline{a}'_2$, we have $D(\overline{a}_i, \overline{a}'_i)$.

Lastly we want to show that $\overline{c}'_{12} = \overline{c}_{12} + 0_{\overline{\phi}'}$. Since $d = \sum_{i=1}^n a_i c + b$, $\overline{e} \le a_2 + a_3 + a_4$ and $\overline{e} \le e \le c$,

$$\overline{c}_{12} = (a_1 c + a_2 c + b)ed = (a_1 c + a_2 c)e + b$$
$$= (a_1 c + a_2 c)(a_1 + a_2)(c_{12} + \overline{e}) + b$$
$$= (a_1 c + a_2 c)(c_{12} + (a_1 + a_2)\overline{e}) + b$$
$$= (a_1 c + a_2 c)(c_{12} + a_2 \overline{e}) + b$$
$$= c_{12}(a_1 c + a_2 c) + a_2 \overline{e} + b$$
$$\le c'_{12} u' + v' = \overline{c}'_{12},$$

where we used $c_{12} \le c'_{12}$, $a_1 c + a_2 c \le u'$ and $a_2 \overline{e} \le b \le v'$ in the last line. Also $\overline{a}_2 \le \overline{a}'_2$ implies $v' \le (\overline{a}_2 + v')\overline{c}'_{12} \le \overline{a}'_2 \overline{c}'_{12} = v'$, and since we already know that $d/\overline{a}_3 + \overline{a}_4 \nearrow \overline{a}'_1 + \overline{a}'_2/v'$ the calculation,

$$\overline{c}'_{12} = (d + v')\overline{c}'_{12} = (\overline{a}_1 + \overline{a}_2 + v')\overline{c}'_{12}$$
$$= (\overline{c}_{12} + \overline{a}_2 + v')\overline{c}'_{12} = \overline{c}_{12} + (\overline{a}_2 + v')\overline{c}'_{12} = \overline{c}_{12} + v'$$

completes this step.

Step 4: As in Step 2 we use Lemma 3.18 to construct a new 4-frame

$$\phi = (a_i, c_{1j}) = \overline{\phi}^u = (\overline{a}_i + u, \overline{c}_{1j} + u),$$

where u is derived from $u_2 = \overline{a}_2(\alpha^p_{\phi}(\overline{a}_1) + \overline{a}_1)$ and $u_1 = \overline{a}_1(u_2 + \overline{c}_{12})$. By Lemma 3.23 ϕ is a 4-frame of characteristic p and as before $f(\phi) = \hat{\phi}$.

Let $w_1' = u_1 + v'$, $w_i' = \overline{a}_i'(w_1' + \overline{c}_{1i}')$, $w' = \sum_{i=1}^4 w_i'$ and consider the 4-frame

$$\phi' = (a_i', c_{1j}') = \overline{\phi}'^{w'} = (\overline{a}_i' + w', \overline{c}_{1j}' + w').$$

Since $\overline{\phi}'$ was of characteristic q, so is ϕ' (Lemma 3.22).

It remains to show that $D(a_i, a_i')$ holds. Note that $1_\phi = 1_{\overline{\phi}} = d$ and $0_{\phi'} = w' \geq w_i \geq v' \geq \overline{a}_3 + \overline{a}_4$. Therefore

$$\begin{aligned} 1_\phi + w' &= \overline{a}_1 + \overline{a}_2 + \overline{a}_3 + \overline{a}_4 + w' \\ &= \overline{a}_1' + \overline{a}_2' + w' = a_1' + a_2'. \end{aligned}$$

Also $u_2 = \overline{a}_2(u_1 + \overline{c}_{12})$ (see proof of Lemma 3.23) and $u_1 = \overline{a}_1(u_2 + \overline{c}_{12})$ imply $u_2 + \overline{c}_{12} = u_1 + \overline{c}_{12}$. Together with $\overline{c}_{12}' = \overline{c}_{12} + v'$ from Step 3 we have

$$\begin{aligned} w_1' + \overline{c}_{12} &= u_1 + \overline{c}_{12} + v' = u_2 + \overline{c}_{12} + v' \\ w_2' &= \overline{a}_2'(w_1' + \overline{c}_{12}') = \overline{a}_2'(u_2 + \overline{c}_{12} + v') \\ &= u_2 + v' + \overline{a}_2'\overline{c}_{12} = u_2 + v' \end{aligned}$$

A last calculation shows that

$$\begin{aligned} 1_\phi w' &= 1_\phi(u_1 + u_2 + w_3' + w_4') \\ &= u_1 + u_2 + 1_\phi(\overline{a}_1' + \overline{a}_2')(w_3' + w_4') \\ &= u_1 + u_2 + 1_\phi v' = u_1 + u_2 + \overline{a}_3 + \overline{a}_4 \\ &= u + \overline{a}_3 + \overline{a}_4 = a_3 + a_4. \end{aligned}$$

Since $\overline{a}_1 \leq \overline{a}_1'$, $\overline{a}_2 \leq \overline{a}_2'$ and $u = 0_\phi \leq w'$ it follows that $a_1 \leq a_1'$ and $a_2 \leq a_2'$. Hence $D(a_i, a_i')$ holds.

Denoting ϕ, ϕ' by $\overline{\phi}, \overline{\phi}'$ and $\hat{\phi}, \hat{\phi}'$ by ϕ, ϕ' we see that condition $(*)$ is now satisfied. $\square$

Note that the lattice L in the preceding theorem has finite length.

Let $\mathcal{M}_{Fl}$ be the class of all modular lattices of finite length, and denote by $\mathcal{M}_Q$ the collection of all subspaces of vector spaces over the rational numbers.

By a result of Herrmann and Huhn [75] $\mathcal{M}_Q \subseteq (\mathcal{M}_F)^{\mathcal{V}}$. Furthermore Herrmann [84] shows that any modular variety that contains $\mathcal{M}_Q$ cannot be both finitely based and generated by its members of finite length. From these results and Freese's Theorem one can obtain the following conclusions.

COROLLARY 3.26

(i) Both $(\mathcal{M}_F)^{\mathcal{V}}$ and $(\mathcal{M}_{Fl})^{\mathcal{V}}$ are not finitely based.

(ii) $(\mathcal{M}_F)^{\mathcal{V}} \subset (\mathcal{M}_{Fl})^{\mathcal{V}} \subset \mathcal{M}$ and all three varieties are distinct.

(iii) The variety of Arguesian lattices is not generated by its members of finite length.

3.4 Covering Relations between Modular Varieties

The structure of the bottom of $\Lambda_{\mathcal{M}}$. In Section 2.1 we saw that the distributive variety $\mathcal{D}$ is covered by exactly two varieties, $\mathcal{M}_3$ and $\mathcal{N}$. The latter is nonmodular, and its covers will be studied in the next chapter. Which varieties cover $\mathcal{M}_3$? Grätzer [66] showed that if a finitely generated modular variety $\mathcal{V}$ properly contains $\mathcal{M}_3$, then $M_4 \in \mathcal{V}$ or $M_{3^2} \in \mathcal{V}$ or both these lattices are in $\mathcal{V}$ (see Figure 2.1 and 3.6). The restriction that $\mathcal{V}$ should be finitely generated was removed by Jónsson [68]. In fact Jónsson showed that for any modular variety $\mathcal{V}$ the condition $M_{3^2} \notin \mathcal{V}$ is equivalent to $\mathcal{V}$ being generated by its members of length ≤ 2. The next few lemmas lead up to the proof of this result.

Recall from Section 1.4 that principal congruences in a modular lattice can be described by sequences of transpositions, which are all bijective. Two nontrivial quotients in a modular lattice are said to be projective to each other if they are connected by some (alternating) sequence of (bijective) transpositions. For example the sequence $a_0/b_0 \nearrow a_1/b_1 \searrow \ldots \nearrow a_n/b_n$ makes a_0/b_0 and a_n/b_n projective to each other in n steps. This sequence is said to be *normal* if $b_k = b_{k-1}b_{k+1}$ for even k and $a_k = a_{k-1} + a_{k+1}$ for odd k $(k = 1, \ldots, n-1)$. It is *strongly normal* if in addition for even k we have $b_{k-1} + b_{k+1} \geq a_k$ and for odd k $a_{k-1}a_{k+1} \leq b_k$.

LEMMA 3.27 (Grätzer [66]). *In a modular lattice any alternating sequence of transpositions can be replaced by a normal sequence of the same length.*

PROOF. Pick any three consecutive quotients from the sequence. By duality we may assume that $a/b \nearrow x/y \searrow c/d$. If this part of the sequence is not normal (i.e. $a + c < x$), then we replace x/y by $a + c/y(a + c)$. To see that $a/b \nearrow a + c/y(a + c)$ we only have to observe that $ay(a + c) = ay = b$ and, by modularity, $a + y(a + c) = (a + y)(a + c) = x(a+c) = a+c$. Similarly $a+c/y(a+c) \searrow c/d$. Notice also that the normality of adjacent parts of the sequence is not disturbed by this procedure, for suppose u/v is the quotient that precedes a/b in the sequence, then $vy = b$ implies $vy(a + c) = b(a + c) = b$. Thus we can replace quotients as necessary, until the sequence is normal. □

Grätzer also observed that the six elements of a normal sequence $a/b \nearrow x/y \searrow c/d$ are generated by a, y and c. Hence, in a modular lattice, they generate a homomorphic image of the lattice in Figure 3.4 (i) (this is the homomorphic image of the free modular lattice $F_{\mathcal{M}}(a, y, c)$ subject to the relations $a + y = a + c = y + c$).

If the sequence is also strongly normal, then $y = y + ac$, and so a, y and c generate a homomorphic image of the lattice in Figure 3.4 (ii). Of course the dual lattices are generated by a (strongly) normal sequence $a/b \searrow x/y \nearrow c/d$. Figure 3.4 (ii) also shows that strongly normal sequences cannot occur in a distributive lattice, unless all the quotients are trivial. However Jónsson [68] proved the following:

LEMMA 3.28 *Suppose L is a modular lattice and p/q and r/s are nontrivial quotients of L that are projective in n steps. If no nontrivial subquotients of p/q and r/s are projective in fewer than n steps, then either $n \leq 2$ or else p/q and r/s are connected by a strongly normal sequence, also in n steps.*

PROOF. Let $p/q = a_0/b_0 \sim a_1/b_1 \sim \ldots \sim a_n/b_n = r/s$ (some $n \geq 3$) be the sequence that connects the two quotients. By Lemma 3.27 we can assume that it is normal. If it is not strongly normal, then for some k with $0 < k < n$ we have $a_{k-1}/b_{k-1} \nearrow a_k/b_k \searrow a_{k+1}/b_{k+1}$,

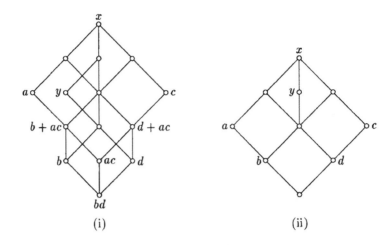

Figure 3.4

but $a_{k_1} a_{k+1} \not\leq b_k$, or dually. Let $c_{k-1} = b_{k-1} + a_{k-1} a_{k+1}$, and for $i \leq n$, $i \neq k-1$ we define c_i to be the element of a_i/b_i that corresponds to c_{k-1} under the given (bijective) transpositions. With reference to Figure 3.4 (i) it is straightforward to verify that

$$c_{k-1}/b_{k-1} \searrow a_{k-1} a_{k+1}/b_{k-1} b_{k+1} \nearrow c_{k+1}/b_{k+1}.$$

Since $0 < k < n$ and $n \geq 3$ we have $k > 1$ or $k < n-1$ (or both). In the first case

$$c_{k-2}/b_{k-2} \searrow a_{k-1} a_{k+1}/b_{k-1} b_{k+1} \nearrow c_{k+1}/b_{k+1}$$

and in the second

$$c_{k-1}/b_{k-1} \searrow a_{k-1} a_{k+1}/b_{k-1} b_{k+1} \nearrow c_{k+2}/b_{k+2}.$$

Either way it follows that the nontrivial subintervals c_0/b_0 of p/q and c_n/b_n of r/s are projective in $n-1$ steps. This however contradicts the assumption of the lemma.   □

LEMMA 3.29 (Jónsson [68]). *Let L be a modular lattice such that M_{3^2} is not a homomorphic image of a sublattice of L. If $(v < x, y, z < u)$ and $(v' < x', y', z' < u')$ are diamonds in L such that $y' = yu'$ and $z = z' + v$, then $u/v \searrow u'/v'$ (refer to Figure 3.5(i)).*

PROOF. Observe firstly that the conditions imply $u/y \searrow z'/v'$, since $y + z' = (y+v) + z' = y + z = u$ and $yz' = y(u'z') = y'z' = v'$. Let $w = v + u'$. Then $w \geq v + z' = z$ and $u \geq y', z'$ imply $u \geq u'$, hence $w \in u/z$. We show that $w = u$ and dually $vu' = v'$, which gives the desired conclusion.

Note that we cannot have $w = z$ since then the two diamonds would generate a sublattice that has M_{3^2} as homomorphic image. So suppose $z < w < u$. Because all the edges of a diamond are projective to another, the six elements w, v, x, y, z, u generate the lattice in Figure 3.5 (ii). Under the transposition $u/y \searrow z'/v'$ the element $xw + y$

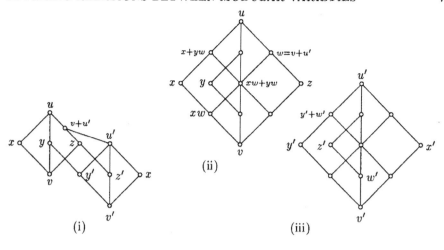

Figure 3.5

is sent to $w' = z'(xw + y)$, and together with v', x', y', z', u' these elements generate the lattice in Figure 3.5 (iii). It is easy to check that $(xw + yw < x + yw, xw + y, w < u)$ and $((x' + w')(y' + w') < y' + w', z' + x'(y' + w'), x' + w' < u')$ are diamonds (they appear in Figure 3.5 (ii) and (iii)). We claim that $w/xw + yw \searrow u'/y' + w'$. Indeed, since $u' \leq w \leq (x + u')(y + u')$ we have

$$xw + yx + u' = u' + xw + u' + yw = (u' + x)w + (u' + y)w = w$$
$$y' + w' = y' + z'(xw + y) = (y' + z')(xw + y)$$
$$= u'(xw + y) = u'(xw + y)w = u'(xw + yw).$$

But this means that the two diamonds form a sublattice of L that is isomorphic to the lattice in Figure 3.5 (iv), and therefore has M_{32} as a homomorphic image. This contradiction shows that we must have $u = w$. □

LEMMA 3.30 (Jónsson [68]). *If L is a modular lattice such that M_{32} is not a homomorphic image of a sublattice of L, then any two quotients in L that are projective to each other have nontrivial subquotients that are projective to each other in three steps or less.*

PROOF. It is enough to prove the theorem for two quotients a/b and c/d that are projective in four steps, since longer sequences can be handled by repeated application of this case. We assume that no nontrivial subquotients of a/b and c/d are projective in less than four steps and derive a contradiction. By Lemma 3.28 there exists a strongly normal sequence of transpositions

$$a/b = a_0/b_0 \nearrow a_1/b_1 \searrow a_2/b_2 \nearrow a_3/b_3 \searrow a_4/b_4 = c/d$$

or dually. Associated with this sequence are three diamonds $(b_0 + b_2 < a_0 + b_2, b_1, b_0 + a_2 < a_1)$, $(b_2 < b_1 a_3, a_2, a_1 b_3 < a_1 a_3)$ and $(b_2 + b_4 < a_2 + b_4, b_3, a_2 + b_4 < a_3)$. The first and the second, and the second and the third diamond satisfy the conditions of Lemma 3.29, since

$$b_1 a_3 = b_1(a_1 a_3), \qquad b_0 + a_2 = a_2 + (b_0 + b_2)$$
$$a_1 b_3 = b_3(a_1 a_3), \qquad a_2 + b_4 = a_2 + (b_0 + b_4)$$

whence we conclude that

$$(*) \qquad a_1/b_0 + b_2 \searrow a_1 a_2/b_2 \nearrow a_3/b_2 + b_4.$$

This enables us to show that a/b and c/d are projective in 2 steps. In fact

$$(**) \qquad a_0/b_0 \nearrow a_2 + b_0 + b_4/a_1 + a_3 \searrow a_4/b_4$$

as is shown by the following calculations

$$
\begin{aligned}
a_0 + (a_2 + b_0 + b_4) &= a_1 + b_4 = a_1 + a_1 a_3 + b_2 + b_4 \qquad (a_1 \ge a_1 a_3 + b_2) \\
&= a_1 + a_3 \qquad (a_1 a_3 + (b_2 + b_4) = a_3 \text{ by } (*)) \\
a_0(a_2 + b_0 + b_4) &= b_0 + a_0(a_2 + b_4) \qquad \text{(by modularity)} \\
&= b_0 + a_0 a_1(a_2 + b_4) \qquad (a_1 \ge a_0) \\
&= b_0 + a_0(a_2 + a_1 b_4) \qquad \text{(by modularity)} \\
&\le b_0 + a_0(a_2 + b_2) = b_0 + a_1 a_2 = b_0,
\end{aligned}
$$

where the inequality holds since $a_1 b_4 = a_1 a_3 b_4 \le a_1 a_3(b_2 + b_4) = b_2$ by $(*)$. The second part of $(**)$ follows by symmetry. Since a/b and c/d were assumed to be projective in not less than 4 steps, this contradiction completes the proof. $\qquad\qquad \square$

LEMMA 3.31 *Suppose L is a modular lattice with $b < a \le d < c$ in L. If a/b and c/d are projective in three steps, then a/b transposes up onto a lower edge of a diamond and c/d transposes down onto an upper edge of a diamond.*

PROOF. Since $a \le d$, no nontrivial subintervals of a/b and c/d are projective to each other in less than three steps. Hence by Lemma 3.29 a/b and c/d are connected by a strongly normal sequence of length 3, say

$$a/b = a_0/b_0 \nearrow a_1/b_1 \searrow a_2/b_2 \nearrow a_3/b_3 = c/d$$

(the dual case cannot apply). Then a/b transposes up onto $a_0 + b_2/b_0 + b_2$ of the diamond $(b_0 + b_2 < a_0 + b_2, b_1, b_0 + a_2 < a_1)$ (see Figure 3.4(ii)) and c/d transposes down onto $a_1 a_3/b_1 b_3$ of $(b_2 < b_1 a_3, a_2, a_1 b_3 < a_1 a_3)$ as required. $\qquad\qquad \square$

THEOREM 3.32 (Jónsson [68]). *For any variety $\mathcal{V}$ of modular lattices the following conditions are equivalent:*

(i) *$M_{3^2} \notin \mathcal{V}$;*

(ii) *every subdirectly irreducible member of $\mathcal{V}$ has dimension two or less;*

(iii) *the inclusion $a(b + cd)(c + d) \le b + ac + ad$ holds in $\mathcal{V}$.*

PROOF. Suppose $M_{3^2} \notin \mathcal{V}$ but some subdirectly irreducible lattice L in $\mathcal{V}$ has dimension greater than two. Then L contains a four element chain $a > b > c > d$. Since L is subdirectly irreducible $\operatorname{con}(a, b)$ and $\operatorname{con}(b, c)$ cannot have trivial intersection, and therefore some nontrivial subquotients a'/b' of a/b and p/q of b/c are projective to each other. By Lemma 3.30 we can assume that they are projective in three steps. Similarly some nontrivial subquotients p'/q' of p/q and c'/d' of c/d are projective to each other, again in three steps. Since all transpositions are bijective, p'/q' is also projective to a subquotient of

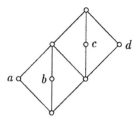

Figure 3.6

a'/b' in three steps. From Lemma 3.31 we infer that p'/q' transposes up onto a lower edge of a diamond and down onto an upper edge of a diamond. It follows that the two diamonds generate a sublattice of L which has M_{3^2} as homomorphic image. This contradicts (i), thus (i) implies (ii).

Every variety is generated by its subdirectly irreducible members, so to prove that (ii) implies (iii), we only have to observe that the inclusion $a(b + cd)(c + d) \leq b + ac + ad$ holds in every lattice of dimension 2. Indeed, in such a lattice we always have $c \leq d$ or $d \leq c$ or $cd = 0$. In the first case $a(b + cd)(c + d) = a(b + c)d \leq ad \leq b + ac + ad$, in the second $a(b + cd)(c + d) = a(b + d)c \leq ac \leq b + ac + ad$ and in the third $a(b + cd)(c + d) = ab(c + d) \leq b \leq b + ac + ad$.

Finally, Figure 3.6 shows that the inclusion fails in M_{3^2}, and therefore (iii) implies (ii).   □

For any cardinal $\alpha \geq 3$ there exists up to isomorphism exactly one lattice M_α with dimension 2 and α atoms (see Figure 2.2). For $\alpha = n \in \omega$ each M_n generates a variety $\mathcal{M}_n$, while for $\alpha \geq \omega$ all the lattices M_α generate the same variety $\mathcal{M}_\omega$ since they all have the same finitely generated sublattices. Clearly $\mathcal{M}_n \subseteq \mathcal{M}_\omega$, and by Jónsson's Lemma $\mathcal{M}_{n+1}$ covers $\mathcal{M}_n$ for $3 \leq n \in \omega$. The above theorem implies that $M_{3^2} \notin \mathcal{M}_n$ for all $n \geq 3$, and conversely, if $\mathcal{V}$ is a variety of modular lattices that satisfies $M_{3^2} \notin \mathcal{V}$, then $\mathcal{V}$ is either $\mathcal{T}$, $\mathcal{D}$, $\mathcal{M}_\omega$ or $\mathcal{M}_n$ for some $n \in \omega$, $n \geq 3$. Thus we obtain:

COROLLARY 3.33 (Jónsson [68]). *In the lattice* Λ, *the variety* $\mathcal{M}_n$ $(3 \leq n \in \omega)$ *is covered by exactly three varieties:* $\mathcal{M}_{n+1}$, $\mathcal{M}_n + \mathcal{M}_{3^2}$ *and* $\mathcal{M}_n + \mathcal{N}$.

PROOF. $\mathcal{M}_n \cap \mathcal{M}_{3^2} = \mathcal{M}_3$ is covered by $\mathcal{M}_{3^2}$, and $\mathcal{M}_n \cap \mathcal{N} = \mathcal{D}$ is covered by $\mathcal{N}$. By the distributivity of Λ, $\mathcal{M}_n \cap \mathcal{M}_{3^2}$ and $\mathcal{M}_n \cap \mathcal{N}$ cover $\mathcal{M}_n$. Suppose a variety $\mathcal{V}$ properly includes $\mathcal{M}_n$. If $\mathcal{V}$ contains a nonmodular lattice, then $\mathcal{N} \in \mathcal{V}$, hence $\mathcal{M}_n + \mathcal{N} \subseteq \mathcal{V}$. If $\mathcal{V}$ contains only modular lattices, then either $M_{3^2} \in \mathcal{V}$ or $M_{3^2} \notin \mathcal{V}$. In the first case we have $\mathcal{M}_n + \mathcal{M}_{3^2} \subseteq \mathcal{V}$, while in the latter case Theorem 3.32 implies that $\mathcal{V} = \mathcal{M}_k$ for some $n < k \in \omega$. Hence $\mathcal{M}_{n+1} \subseteq \mathcal{V}$, and the proof is complete.   □

The proof in fact shows that, for $n \geq 3$, $C(\mathcal{M}_n) = \{\mathcal{M}_{n+1}, \mathcal{M}_n + \mathcal{M}_{3^2}, \mathcal{M}_n + \mathcal{N}\}$ strongly covers $\mathcal{M}_n$ (see Section 2.1). But this is to be expected in view of Theorem 2.2 and the result that every finitely generated lattice variety is finitely based (Section 5.1).

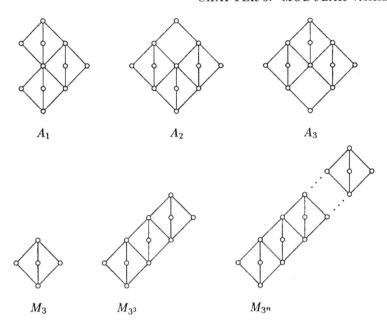

Figure 3.7

Observe that $\mathcal{M}_3$ has two join irreducible covers ($\mathcal{M}_4$ and $\mathcal{M}_{3^2}$) whereas $\mathcal{M}_n$ ($4 \leq n \in \omega$) has only one.

Further results on modular varieties. Consider the lattices in Figure 3.7. The main result of Hong [**72**] is the following:

THEOREM 3.34 *Let L be a subdirectly irreducible modular lattice and suppose*

$$A_1, A_2, A_3, M_{3^n} \notin \mathbf{HS}\{L\}.$$

Then the dimension of L is less than or equal to n.

The proof of this theorem is based on a detailed analysis of how the diamonds that are associated with a normal sequence of quotients fit together.

We list some consequences of this result. Let $\mathcal{M}_{3^n}$, $\mathcal{A}_1$, $\mathcal{A}_2$, $\mathcal{A}_3$ and $\mathcal{P}_2$ be the varieties generated by the lattices M_{3^2}, A_1, A_2, A_3 and P_2 respectively.

COROLLARY 3.35 (Hong [**70**]). *For $2 \leq n \in \omega$ the variety $\mathcal{M}_{3^n}$ is covered by the varieties* $\mathcal{M}_{3^{n+1}}$, $\mathcal{M}_{3^n} + \mathcal{M}_4$, $\mathcal{M}_{3^n} + \mathcal{A}_1$, $\mathcal{M}_{3^n} + \mathcal{A}_2$, $\mathcal{M}_{3^n} + \mathcal{A}_3$, $\mathcal{M}_{3^n} + \mathcal{P}_2$, $\mathcal{M}_{3^n} + \mathcal{N}$.

Let $\mathcal{M}_n^m$ be the variety generated by all modular lattices whose length does not exceed m and whose width does not exceed n ($1 \leq m, n \leq \infty$). Note that Theorem 3.32 implies $\mathcal{M}_\infty^2 = \mathcal{M}_\omega$, and since every lattice of length at most 3 can be embedded in the subspace lattice of a projective plane ([**GLT**] p.214), $\mathcal{M}_\infty^3$ is the variety generated by all such subspace lattices.

COROLLARY 3.36 (Hong [**72**]). *The variety $\mathcal{M}^3_\infty$ is strongly covered by the collection*

$$\{\mathcal{M}^3_\infty + \mathcal{M}_{3^2}, \ M^3_\infty + \mathcal{A}_1, \ M^3_\infty + \mathcal{A}_2, \ M^3_\infty + \mathcal{A}_3, \ M^3_\infty + \mathcal{N}\}$$

From Theorem 2.2 one may now deduce that $\mathcal{M}^3_\infty$ is finitely based.

Considering the varieties $\mathcal{M}^\infty_n$, we first of all note that since M_3 has width 3, $\mathcal{M}^\infty_1$ and $\mathcal{M}^\infty_2$ are both equal to the distributive variety $\mathcal{D}$. The two modular varieties which cover $\mathcal{M}_3$ are generated by modular lattices of width 4, hence $\mathcal{M}^\infty_3 = \mathcal{M}_3$.

The variety $\mathcal{M}^\infty_4$ is investigated in Freese [**77**]. It is not finitely generated since it contains simple lattices of arbitrary length (Figure 3.8 (i)). Freese obtains the following result:

THEOREM 3.37 *The variety $\mathcal{M}^\infty_4$ is strongly covered by the following collection of ten varieties:*

$$\{\mathcal{M}^\infty_4 + \{L\}^{\mathcal{V}} : L = A_2, A_3, \ldots, A_8, M_5, P_2, N\}.$$

He also gives a complete list of the subdirectly irreducible members of this variety, and shows that it has uncountably many subvarieties. Further remarks about the varieties $\mathcal{M}^m_n$ appear at the end of Chapter 5.

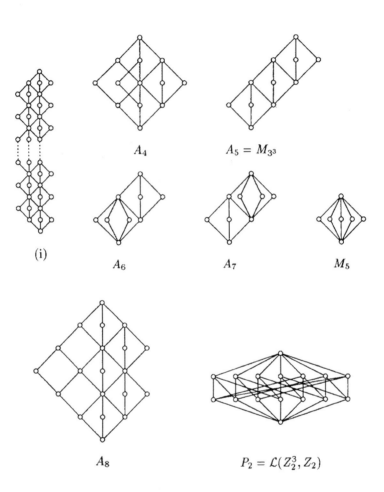

Figure 3.8

Chapter 4

Nonmodular Varieties

4.1 Introduction

The first significant results specifically about nonmodular varieties appear in a paper by McKenzie [72], although earlier studies by Jónsson concerning sublattices of free lattices contributed to some of the results in this paper (see also Kostinsky [72], Jónsson and Nation [75]). Splitting lattices are characterized as subdirectly irreducible bounded homomorphic images of finitely generated free lattices, and an effective procedure for deciding if a lattice is splitting, and to find its conjugate equation (see Section 2.3) is given. Also included in McKenzie's paper are several problems which stimulated a lot of research in this direction. One of these problems was solved when Day [77] showed that the class of all splitting lattices generates the variety of all lattices (Section 2.3).

McKenzie [72] also lists fifteen subdirectly irreducible lattices $L_1, L_2, \ldots, L_{15}$, (see Figure 2.2) each of which generates a join irreducible variety that covers the smallest nonmodular variety $\mathcal{N}$. Davey, Poguntke and Rival [75] proved that a variety, generated by a lattice which satisfies the double chain condition, is semidistributive if and only if it does not contain one of the lattices $M_3, L_1, \ldots, L_5$. Jónsson proved the same result without the double chain condition restriction, and in Jónsson and Rival [79] this is used to show that McKenzie's list of join irreducible covers of $\mathcal{N}$ is complete.

Further results in this direction by Rose [84] prove that there are eight chains of semidistributive varieties, each generated by a finite subdirectly irreducible lattice L_6^n, L_7^n, L_8^n, L_9^n, L_{10}^n, L_{13}^n, L_{14}^n, L_{15}^n ($n \geq 0$, see Figure 2.2), such that $L_i^0 = L_i$, and $\{L_i^{n+1}\}^{\mathcal{V}}$ is the only join irreducible cover of $\{L_i^n\}^{\mathcal{V}}$ for $i = 6, 7, 8, 9, 10, 13, 14, 15$.

Extending some results of Rose, Lee [85] gives a fairly complete description of all the varieties which do not contain any of $M_3, L_2, L_3, \ldots, L_{12}$. In particular, these varieties turn out to be locally finite.

Ruckelshausen [78] obtained some partial results about the covers of $\mathcal{M}_3 + \mathcal{N}$, and Nation [85] [86] has developed another approach to finding the covers of finitely generated varieties, which he uses to show that $\{L_1\}^{\mathcal{V}}$ has ten join irreducible covers, and that above $\{L_{12}\}^{\mathcal{V}}$ there are exactly two join irreducible covering chains of varieties. These results are mentioned again in more detail at the end of Section 4.4.

The notions of splitting lattices and bounded homomorphic images have been discussed in Section 2.3, so this chapter covers the results of Jónsson and Rival [79], Rose [84] and Lee [85].

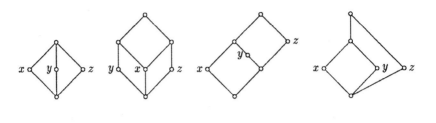

$$M_3(x,y,z) \qquad\qquad L_1(x,y,z) \qquad\qquad L_3(x,y,z) \qquad\qquad L_4(x,y,z)$$
$$(L_2 \text{ is dual}) \qquad\qquad\qquad\qquad (L_5 \text{ is dual})$$

$$x+y = y+z \qquad\quad y(x+z) = xz \qquad\quad x+z = y+z \qquad\quad x+z = y+z$$
$$= x+z \qquad\qquad (x+y)z = xy \qquad\qquad xy = xz \qquad\qquad (x+y)z = xy$$
$$xy = yz = xz \qquad (x+y)(x+z) = x \qquad x+yz = x+y$$
$$x \le y+z \qquad\qquad (x+y)z = yz$$

Figure 4.1

4.2 Semidistributivity

Recall from Section 2.3 that a lattice L is *semidistributive* if for any $u,v,x,y,z \in L$

$$(\text{SD}^+) \qquad u = x+y = x+z \quad \text{implies} \quad u = x+yz \qquad \text{and dually}$$
$$(\text{SD}^\cdot) \qquad u = xy = xz \qquad\quad \text{implies} \quad u = x(y+z).$$

A glance at Figure 4.1 shows that the lattices M_3, L_1, L_2, L_3, L_4 and L_5 fail to be semidistributive and hence they cannot be a sublattice of any semidistributive lattice. The next lemma implies that for finite lattices the converse is also true. Given a lattice L and three *noncomparable* elements $x,y,z \in L$ we will write $L_i(x,y,z)$ to indicate that these elements generate a sublattice of L isomorphic to L_i, $i = 1,2,3,4,5$ (Figure 4.1). Algebraically this is verified by checking that the corresponding defining relations (below Figure 4.1) hold.

We denote by $\mathcal{I}L$ and $\mathcal{F}L$ the ideal and filter lattice of L respectively ($\mathcal{F}L$ is ordered by reverse inclusion). L is embedded in $\mathcal{I}L$ via the map $x \mapsto (x]$ and in $\mathcal{F}L$ via $x \mapsto [x)$. We identify L with its image in $\mathcal{I}L$ and $\mathcal{F}L$. Of course $\mathcal{I}L$ ($\mathcal{F}L$) is (dually) algebraic with the (dually) compact elements being the principal ideals (filters) of L. Hence both lattices are weakly atomic (i.e. in any quotient u/v we can find $r,s \in u/v$ such that $r \succ s$). In particular, given $a,b \in L$, there exists $c \in \mathcal{I}L$ satisfying $a \le c \prec a+b$.

Note also that $\mathcal{I}L$ is *upper continuous*, i.e. for any $x \in \mathcal{I}L$ and any chain $C \subseteq \mathcal{I}L$, we have $x \sum C = \sum_{y \in C} xy$ (see [**ATL**] p. 15).

LEMMA 4.1 (Jónsson and Rival [**79**]). *If a lattice L is not semidistributive, then either $\mathcal{I}\mathcal{F}\mathcal{I}L$ or $\mathcal{F}\mathcal{I}\mathcal{F}L$ contains a sublattice isomorphic to one of the lattices M_3, L_1, L_2, L_3, L_4 or L_5.*

PROOF. Suppose that L is not semidistributive. By duality we may assume that there

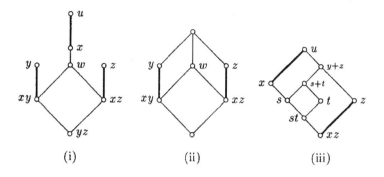

Figure 4.2

exist $u, x, y, z \in L$ such that

$$(*) \qquad u = x + y = x + z \quad \text{but} \quad x + yz < u.$$

As a first observation we have that x, y and z must be noncomparable. By the weak atomicity of $\mathcal{I}L$, we can find $x' \in \mathcal{I}L$ such that $x + yz \le x' \prec u$, whence it follows that

$$u = x' + y = x' + z, \quad yz \le x' \prec u.$$

In $\mathcal{FI}L$ we can then find minimal elements y', z' subject to the conditions $u = x' + y' = x' + z'$, $y' \le y$ and $z' \le z$. Now $x'y' < y'$ since equality would imply $x' = u$. Furthermore, if $x'y' < w \le y'$, then $x' < x' + w$ (equality would imply $w \le x'y'$) and $x' + w \le x' + y' = u$. Hence $x' + w = u$ and by the minimality of y', $w = y'$. It follows that y' covers $x'y'$, and similarly z' covers $x'z'$. So, dropping the primes, we have found $u, x, y, z \in \mathcal{FI}L$ satisfying $(*)$ and

$$(**) \qquad z \le x \prec u, \quad xy \prec y, \quad xz \prec z \quad \text{see Figure 4.2 (i).}$$

Since $xy \prec y$, we have either $y(xy+z) = xy$ or $y \le xy + z$, and similarly $z(xz+y) = xz$ or $z \le xz + y$. We will show that in each of the four cases that arise, the lattice $\mathcal{FI}L$ or $\mathcal{IFI}L$ must contain M_3 or one of the L_i $(i = 1, \ldots, 5)$ as a sublattice.

Case 1: $y \le xy + z$ and $z \le xz + y$. Since x, y and z are noncomparable, so are xy and xz ($xy \le xz$ would imply $y \le xy + z \le xz + z = z$). Let $w = xy + xz$, then $y, z \not\le w \le x$ and $w \not\le y, z$ since $xy \prec y$ and $xz \prec z$. The following calculations show that we in fact have $L_2(w, y, z)$: $wy + wz = xy + xz = w$; $w \ge xyz = yz$; $y + xz \le y + z$ and equality follows from the assumption that $y + xz \ge z$; similarly $z + xy = y + z$ (see Figure 4.1 (ii), 4.2 (i) and (ii)).

Case 2: $y \le xy + z$ and $z(xz + y) = xz$. Let $s = x(y + z)$ and $t = xz + y$. The most general relationship between x, s, t and z is pictured in Figure 4.2 (iii). We will show that either $L_5(s, t, z)$ or $L_3(z, s + t, x)$ (see Figure 4.1). Clearly $sz = x(y+z)z = xz = xt$ by assumption. Furthermore t and z are noncomparable ($z \not\le t$ since $tz = xz < z$, $y \le t \not\le z$ since $y \not\le z$), as are s and z ($z \not\le s \le x$, $s \not\le z$ else $s + z = z = y + z$), and $t \not\le s$ since $y \le t$, $s \le x$ but $y \not\le x$. Suppose now that $s + t = y + z$. Then $s \not\le t$ (else $y + z = s + t = t \ge z$,

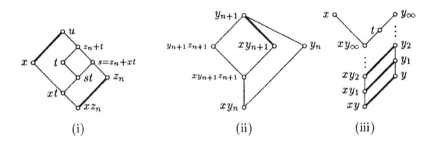

Figure 4.3

contradicting $t \not\geq z$) and therefore s, t, z are noncomparable. Also $st = xt$ since $t \leq y + z$, thus $st + z = xt + z \geq xy + z \geq y + z$ (by assumption), whence $st + z = y + z$. This shows $L_5(s, t, z)$. On the other hand $s + t < y + z$ implies that $z, s + t, x$ are noncomparable, and so $L_3(z, s + t, x)$ follows from the calculations:

$$
\begin{aligned}
u = z + x &= s + t + x & \text{(since } x \prec u) \\
z(s + t) &= zx & \text{(since } zx \prec x) \\
z + x(s + t) &= z + s = z + (s + t) \\
(z + s + t)x &= (y + z)x = s = (s + t)x.
\end{aligned}
$$

Case 3: $y(xy + z) = xy$ and $z \leq xz + y$. This case is symmetric to the preceding case.

Case 4: $y(xy + z) = xy$ and $z(xz + y) = xz$. We claim that for $n = 0, 1, \ldots$ one can find increasing chains of elements $y_n \in y + z/y$ and $z_n \in y + z/z$ such that (*) and (**) hold with y and z replaced by y_n and z_n. Indeed, let $y_0 = y$, $z_0 = z$ and

$$ y_{n+1} = y_n + xz_n, \qquad z_{n+1} = z_n + xy_n. $$

Then $y_0 \leq y + z$ and $z_0 \leq y + z$ and if we suppose that $y_n, z_n \leq y + z$ then clearly $y_{n+1} = y_n + xz_n \leq y + z$ and $z_{n+1} \leq y + z$. Now suppose that (*) and (**) hold with y and z replaced by y_n and z_n for some $n \geq 0$. We show that the same is true for y_{n+1} and z_{n+1}. Firstly $x + y_{n+1} = x + y_n + xz_n = x + y_n = u$ by hypothesis, and similarly $x + z_{n+1} = u$. Further we may assume that $y_n z_{n+1} = xy_n$ and $z_n y_{n+1} = xz_n$, for otherwise one of the three previous cases would apply. Now $z \leq z_{n+1}$ and $z \not\leq x$, so $z_{n+1} \not\leq x$ and hence $xz_{n+1} < z_{n+1}$. If $xz_{n+1} < t < z_{n+1}$ then put $s = z_n + xt$ (Figure 4.3 (i)). We show that either $L_3(z_n, t, x)$ or $L_4(s, t, x)$ or $L_3(z_n, st, x)$, from which it follows that we may assume $xz_{n+1} \prec z_{n+1}$. $u = z_n + x = t + x$ since $x \prec u$, and $xz_n = tz_n$ since $xz_n \prec z_n$. Also $xt \leq x(z_n + t) \leq xz_{n+1} \leq xt$ shows that $xt = x(z_n + t) = xz_{n+1}$ and x, t, z_n are noncomparable.

Now either $z_n + t = s$, which implies $L_3(z_n, t, x)$, or $z_n + t > s$ in which case we have $L_4(s, t, x)$ (if $st = xt$) or $L_3(z_n, st, x)$ (if $st > xt$). Similarly we may assume that $xy_{n+1} \prec y_{n+1}$. Finally we can assume that $y_{n+1}z_{n+1} \leq x$, otherwise we obtain $L_5(y_{n+1}z_{n+1}, xy_{n+1}, y_n)$ (Figure 4.3 (ii)).

In $\mathcal{IFIL}$ we now form the join y_∞ of all the y_n and the join z_∞ of all the z_n. Clearly

$$u = x + y_\infty = x + z_\infty, \qquad y_\infty + z_\infty = y + z.$$

Furthermore, $x \not\leq y_\infty$ since x is compact and $x \not\leq y_n$ for all n. Therefore $xy_\infty < y_\infty$ and if $xy_\infty \leq t < y_\infty$ then there exists $m \in \omega$ such that for all $n \geq m$ $t \not\geq y_n$, hence $ty_n = xy_n$ (Figure 4.3 (iii)). We compute

$$t = ty_\infty = \sum_{n \geq m} ty_n = \sum_{n \geq m} xy_n = xy_\infty$$

where the second and last equality make use of the upper continuity of $\mathcal{IFIL}$. Thus $xy_\infty \prec y_\infty$ and similarly $xz_\infty \prec z_\infty$. Also, for each m, $y_m z_\infty = \sum_{n \in \omega} y_m z_n \leq x$, hence $y_\infty z_\infty \leq x$. Lastly, $xy_n \leq xz_{n+1}$ implies

$$xy_\infty = \sum xy_n \leq \sum xz_{n+1} = xz_\infty$$

and similarly $xz_\infty \leq xy_\infty$. Consequently $xy_\infty = xz_\infty = y_\infty z_\infty$.

Dropping the subscripts we now have $u, x, y, z \in \mathcal{IFIL}$ satisfying (*), (**) and $xy = xz = yz$. Let $t = x(y + z)$. If $t = yz$ then $L_4(y, z, x)$ holds, and if $t > yz$ then we consider the four cases depending on whether or not the equations $y + z = y + t$ and $y + z = t + z$ hold. If both hold, then we get $M_3(w, y, z)$, if both fail then we let $s = (y + w)(w + z)$ to obtain $L_1(s, y, z)$ (here we use $xy \prec y$, $xz \prec z$, see Figure 4.3 (iv)), and if just one equation holds, say $y + t < z + t = y + z$, then $L_4(y, t, z)$ follows. This completes the proof. $\qquad\square$

Semidistributive varieties. If L is a finite lattice, then $\mathcal{IFIL} \cong \mathcal{FIFL} \cong L$, so L is semidistributive if and only if L excludes M_3, L_1, L_2, L_3, L_4 and L_5. We say that a *variety* $\mathcal{V}$ *of lattices is semidistributive* if every member of $\mathcal{V}$ is semidistributive. The next theorem characterizes all the semidistributive varieties.

THEOREM 4.2 (Jónsson and Rival [79]). *For a given lattice variety $\mathcal{V}$, the following statements are equivalent:*

(i) $\mathcal{V}$ *is semidistributive.*

(ii) $M_3, L_1, L_2, L_3, L_4, L_5 \notin \mathcal{V}$.

(iii) *Both the filter and ideal lattice of $F_\mathcal{V}(3)$ are semidistributive.*

(iv) *Let $y_0 = y$, $z_0 = z$ and, for $n \in \omega$ let $y_{n+1} = y + xz_n$ and $z_{n+1} = z + xy_n$. Then for some natural number m the identity*

$$(\text{SD}'_m) \qquad x(y + z) = xy_m$$

and its dual (SD^+_m) hold in $\mathcal{V}$.

PROOF. Since each of the lattices in (ii) fail to be semidistributive, (i) implies (ii), and (ii) implies (i) follows from Lemma 4.1 and the fact that $L \in \mathcal{V}$ implies $\mathcal{IFIL}, \mathcal{FIFL} \in \mathcal{V}$. Also (i) implies (iii) since $\mathcal{IF}_\mathcal{V}(3), \mathcal{FF}_\mathcal{V}(3) \in \mathcal{V}$.

(iii) implies (iv): By duality it suffices to show that, for some m, $x(y + z) = xy_m$ in the free lattice $F_\mathcal{V}(3)$ of $\mathcal{V}$ generated by x, y, z. By induction one easily sees that $y_n \leq y_{n+1}$

and $z_n \leq z_{n+1}$. In $\mathcal{I}F_V(3)$ we define y_∞ as the join of all the y_n and z_∞ as the join of all the z_n. Now $xy_n \leq xz_{n+1}$ and $xz_n \leq xy_{n+1}$, hence by the upper continuity of $\mathcal{I}F_V(3)$, $xy_\infty = xz_\infty = v$, say. Also $y_n + z_n = y + z$ for each n implies $y_\infty = z_\infty = y + z$. By semidistributivity we therefore have $v = x(y_\infty + z_\infty) = x(y + z)$. Hence $v = \sum_{n \in \omega} xy_n$ is a compact element of $\mathcal{I}F_V(3)$, so for some $m \in \omega$, $x(y + z) = xy_m$.

(iv) implies (i): If $L \in V$ is not semidistributive, then there are elements x, y, z in L such that $xy = xz < x(y + z)$ or dually. Then, for all n, $y_n = y$ and $z_n = z$, whence $xy_n < x(y + z)$. Consequently the identity fails for each n.   □

The fourth statement shows that semidistributivity cannot be characterized by a set of identities, and so the class of all semidistributive lattices does not form a variety.

Semidistributivity and weak transpositions. For the notions of weak projectivity we refer the reader to Section 1.4. The next result concerns the possibility of shortening a sequence of weak transpositions. Suppose in some lattice L a quotient x_0/y_0 projects weakly onto another quotient x_n/y_n in $n > 2$ steps, say

$$x_0/y_0 \nearrow_w x_1/y_1 \searrow_w x_2/y_2 \nearrow_w \ldots x_n/y_n.$$

If there exists a quotient u/v such that

$$x_0/y_0 \searrow_w u/v \nearrow_w x_2/y_2,$$

then we can shorten the sequence of weak transpositions by replacing the quotients x_1/y_1 and x_2/y_2 by the single quotient u/v. In a distributive lattice this can always be done, since we may take $u/v = x_0x_2/y_0y_2$. The nonexistence of such a quotient u/v is therefore connected with the presence of a diamond or a pentagon as a sublattice of L. If L is semidistributive, then this sublattice must of course be a pentagon. The aim of Lemma 4.3 is to describe the location of the pentagon relative to the quotients x_i/y_i.

We introduce the following terminology and notation: A quotient c/a in a lattice L is said to be an N-quotient if there exists $b \in L$ such that $a + b = a + c$ and $ab = ac$. In this case $a, b, c \in L$ generate a sublattice isomorphic to the pentagon N, a condition which we abbreviate by writing $N(c/a, b)$.

LEMMA 4.3 (Jónsson and Rival [79]). *Let L be a semidistributive lattice and suppose $x_0/y_0 \nearrow_w x_1/y_1 \searrow_w x_2/y_2$ in L. Then either*

(i) *there exist $a, b, c \in L$ with $N(c/a, b)$, and b/bc is a subquotient of x_0/y_0 or*

(ii) *there exist $a, b, c, t \in L$ with $N(c/a, b)$, $y_0 < t \leq x_0$ and $t/y_0 \nearrow_w a + b/b$ or*

(iii) *there exists a subquotient p/q of x_0/y_0 such that $p/q \searrow u/v \nearrow x_2/y_2$ for some quotient u/v.*

PROOF. Let $x_0' = x_0(y_1 + y_2)$. (i) If $x_0' + y_1 < x_2 + y_1$, then the elements $a = x_0' + y_1$, $b = x_0$ and $c = y_1 + y_2$ give $N(c/a, b)$ and $b/bc = x_0/x_0' \subseteq x_0/y_0$ (Figure 4.4 (i)).

(ii) Suppose $x_0' + y_1 = x_2 + y_1$. By the semidistributivity of L, $x_2 + y_1 = x_0'x_2 + y_1 = x_0x_2 + y_1$, hence $(x_0x_2 + y_2) + y_1 = x_2 + y_1$. If $x_0x_2 + y_2 < x_2$, then the elements $a = x_0x_2 + y_2$, $b = y_1$, $c = x_2$ satisfy $n(c/a, b)$, and $a + b/b$ transposes down onto the subquotient x_0'/x_0y_1 of x_0/y_0 (Figure 4.4 (ii)).

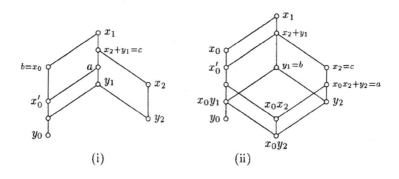

Figure 4.4

(iii) If $x_0x_2 + y_2 = x_2$, then

$$x_0/y_0 \supseteq p/q = x_0y_1 + x_0x_2/x_0y_1 \searrow x_0x_2/x_0y_2 \nearrow x_2/y_2.$$

□

LEMMA 4.4 (Rose [84]). *If L is a semidistributive lattice and $x_0/y_0 \nearrow_\beta x_1/y_1 \searrow_\beta x_2/y_2$ in L, then $x_0/y_0 \searrow x_0x_2/y_0y_2 \nearrow x_2/y_2$.*

PROOF. Since we are dealing with transpositions, $y_0 = x_0y_1$, $y_2 = x_2y_1$ and $x_1 = y_1 + x_0 = y_1 + x_1$. By semidistributivity $x_1 = y_1 + x_0x_2$. Now the bijectivity of the transpositions implies $y_0 + x_0x_2 = (y_0 + x_0x_2 + y_2)x_0 = x_1x_0 = x_0$, and similarly $y_2 + x_0x_2 = x_2$. Also $y_0(x_0x_2) = x_0y_1x_2 = y_0y_1$ and $y_2(x_0x_2) = y_0y_1$. □

3-generated semidistributive lattices. Let F_1, F_2, F_3, F_4 be the lattices in Figure 4.5. It is easy to check that each of these lattices is freely generated by the elements x, y, z subject to the defining relations listed below.

LEMMA 4.5 (Jónsson and Rival [79], Rose [84]). *Let L be a semidistributive lattice generated by the three x, y, z with $x \le xy + z$ and $xz \le y$.*

(i) *If L excludes L_{12} then $L \in \mathbf{H}F_1$.*

(ii) *If L excludes L_{12} and L_7 then $L \in \mathbf{H}F_2$.*

(iii) *If L excludes L_{12} and L_8 then $L \in \mathbf{H}F_3$.*

(iv) *If L excludes L_{12}, L_7 and L_8 then $L \in \mathbf{H}F_4$.*

PROOF. (i) $x \le xy + z$ is equivalent to $xy + z = x + z$, so it suffices to show that under the above assumptions $(x + y)z = yz$. The free lattice determined by the elements x, z, xy, yz and the defining relations $x \le xy + z$ and $xz \le y$ is pictured in Figure 4.6(i). Let $y_0 = x + yz$, $y_1 = xy + y_0z$, $y_2 = yz + xy_1$ and $w = xy + yz$. To avoid L_{12} we must have $y_1 = y_2$.

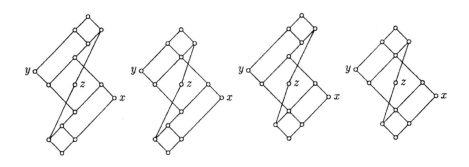

$$F_1$$
$$(x + y)z = yz$$
$$xy + z = x + z$$

$$F_2$$
$$(x + y)z = yz$$
$$xy + z = x + z$$

$$x + y(x + z)$$
$$= (x + y)(x + z)$$

$$F_3$$
$$(x + y)z = yz$$
$$xy + z = x + z$$
$$(x + yz)y$$
$$= xy + yz$$

$$F_4$$
$$(x + y)z = yz$$
$$xy + z = x + z$$
$$(x + yz)y$$
$$= xy + yz$$
$$x + y(x + z)$$
$$= (x + y)(x + z)$$

Figure 4.5

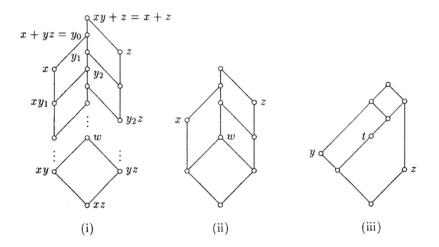

(i)   (ii)   (iii)

Figure 4.6

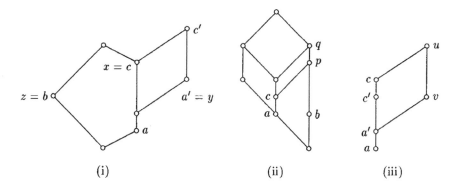

Figure 4.7

Since $w + y_0 z = y_1 = y_2 = w + x y_1$, semidistributivity implies $y_1 = y_2 = w y_0 z x y_1 = w$. Further we compute $y_0 z = y_1 z = w z = y z$. Again by semidistributivity $yz = (y_0 + y)z = (x + yz + y)z = (x + y)z$, as required.

(ii) Let $s = (x + y)(y + z)$ and $t = x + y(x + z)$, then the sublattice of F_1 generated by y, z and t is isomorphic to L_7 (Figure 4.6 (iii)) with critical quotient s/t.

(iii) is dual to (ii), and (iv) follows from (ii) and (iii). □

LEMMA 4.6 (Rose [84]). *Let L be a semidistributive lattice that excludes L_{11} and L_{12}. If $a, b, c, u, v \in L$ with $N(u/v, b)$, $a < c$ and u/v projects weakly onto c/a, then $N(c/a, b)$.*

PROOF. We show that $u/v \nearrow_w c/a$ implies $N(c/a, b)$, then the result follows by repeated application of this result and its dual. Let $x = u$, $y = a$ and $z = b$ (Figure 4.7 (i)). Then $x \le xy + z$ and $xz \le y$, hence by Lemma 4.5 x, y, z generate a homomorphic image of F_1 (Figure 4.5). Computing in this lattice, we have $bc = z(x + y) = zy = ba$ and $b + a = z + y = z + (x + y) = b + c$, which implies $N(c/a, b)$. □

COROLLARY 4.7 *Suppose u/v and c/a are nontrivial quotients in a semidistributive lattice L that excludes L_{11} and L_{12}. If $b \in L$ and $(a, c) \in \text{con}(u, v)$, then $N(u/v, b)$ implies $N(c/a, b)$.*

PROOF. By Lemma 1.11 there is a sequence $a = e_0 < e_1 < \ldots < e_n = c$ such that u/v projects weakly onto e_i/e_{i-1} for each $i = 1, \ldots, n$. By Lemma 4.6 $N(u/v)$ implies $N(e_i/e_{i-1}, b)$, hence $e_i b < e_{i-1}$ and $e_{i-1} + b > e_i$. It follows that $ab = e_1 b = \ldots = e_n b = cb$ and $a + b = c + b$, whence $N(c/a, b)$. □

Figure 4.7 (ii) shows that the above result does not hold if L includes L_{11} or (by duality) L_{12}. The next result shows just how useful the preceding few lemmas are.

THEOREM 4.8 (Rose [84]). *Let L be a subdirectly irreducible semidistributive lattice that excludes L_{11} and L_{12}. Then L has a unique critical quotient.*

PROOF. Suppose to the contrary that c/a and p/q are two distinct critical quotients of L. Then $(p, q) \in \text{con}(a, c)$, hence by Lemma 1.11 there exists $p' \in p/q$ such that $p' > q$

and c/a projects weakly onto p'/q in k steps. We may assume that $c/a \not\subseteq p/q$ (else p/c or a/q is critical and can replace p/q), and $p/q \not\subseteq c/a$. Consequently $k \geq 1$, $p'/q \not\subseteq c/a$ and therefore we can find a nontrivial quotient $u/v \not\subseteq c/a$ such that $c/a \sim_w u/v$. By duality, suppose that $c/a \nearrow_w u/v$, and put $a' = cv$. Since c/a' is also critical, we get $(a', c) \in \mathrm{con}(a', v)$. Again by Lemma 1.11 there exists $c' \in c/a$, $c' > a'$ such that v/a' projects weakly onto c'/a' (see Figure 4.7 (iii)). Consider a shortest sequence

$$v/a' = x_0/y_0 \sim_w x_1/y_1 \sim_w \ldots \sim_w x_n/y_n = c'/a'.$$

Clearly $n \geq 2$ since $c' \not\leq v$. Observe also that if $n = 2$, then we cannot have $v/a' \searrow_w$ $x_1/y_1 \nearrow_w c'/a'$, since that would imply $c' = a' + x_1 \leq v$.

First suppose that $v/a' \nearrow_w x_1/y_1 \searrow_w x_2/y_2$. Then only (i) or (ii) of Lemma 4.2 can apply, since the sequence cannot be shortened if $n \geq 3$, and for $n = 2$ this follows from the observation above. If (i) holds, then there exist $a'', b, c'' \in L$ such that $N(c''/a'', b)$ and $b/bc'' \subseteq v/a'$. Since c'/a' is critical, $(a', c') \in \mathrm{con}(a'', c'')$, whence by Corollary 4.7 we have $N(c'/a', b)$. If (ii) holds, then there exist $a'', b, c'', t \in L$ such that $N(c''/a'', b)$, $t/a'' \subseteq v/a'$ and $t/a' \nearrow_w a'' + b/b$. Again we get $N(c'/a', b)$ from Corollary 4.7. But in both cases we also have $b \leq a'$, so this is a contradiction.

Now suppose that $v/a' \searrow_w x_1/y_1 \nearrow_w x_2/y_2$. As we already noted, this implies $n \geq 3$, so we may only apply the dual parts of (i) or (ii) of Lemma 4.2. That is, there exist $a'', b, c'', t \in L$ with $N(c''/a'', b)$ and either $a'' + b/b \subseteq v/a'$ or $v/t \subseteq v/a'$, $v/t \searrow_w b/bc''$. Again Corollary 4.7 gives $N(c'/a', b)$. In the first case this contradicts $b \geq a'$, and in the second, since $b/bc'' \nearrow v/t$, we have $v = b + t \geq b + a' \geq c'$, and this contradicts $a' = vc'$.□

Notice that if a lattice has a unique critical quotient c/a, then this quotient is prime (i.e. c covers a), c is join irreducible, a is meet irreducible, and $\mathrm{con}(a, c)$ identifies no two distinct elements except c and a. To get a feeling for the above theorem, the reader should check that the lattices $N, L_6, L_7, L_8, L_9, L_{10}, L_{13}, L_{14}$ and L_{15} each have a unique critical quotient, where as L_{11} and L_{12} each have two.

4.3   Almost Distributive Varieties

Recall the definition of the identities $(\mathrm{SD}_m^{\cdot})$ and (SD_m^+) in Theorem 4.2. Of course $(\mathrm{SD}_0^{\cdot})$ and (SD_0^+) only hold in the trivial variety, while

$$(\mathrm{SD}_1^{\cdot}) \qquad x(y + z) = x(y + xz)$$

holds in the distributive variety, but fails in M_3 and N (Figure 4.8). Thus $(\mathrm{SD}_1^{\cdot})$ (and by duality (SD_1^+)) is equivalent to the distributive identity. The first identities that are of interest are therefore

$$(\mathrm{SD}_2^{\cdot}) \qquad x(y + z) = x(y + x(z + xy))$$
$$(\mathrm{SD}_2^+) \qquad x + yz = x + y(x + z(x + y)).$$

Neardistributive lattices. A lattice, or a lattice variety, is said to be *neardistributive* if it satisfies the identities $(\mathrm{SD}_2^{\cdot})$ and (SD_2^+). This definition appears in Lee [85]. By Theorem 4.2 every neardistributive lattice is semidistributive, and it is not difficult (though somewhat tedious) to check that $N, L_6, \ldots, L_{10}, L_{13}, L_{14}$ and L_{15} are all neardistributive.

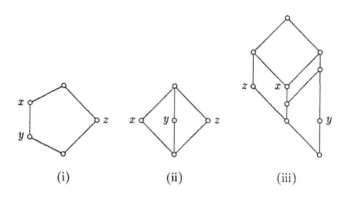

Figure 4.8

On the other hand Figure 4.8 (iii) shows that (SD_2^-) fails in L_{11}, and by duality (SD_2^+) fails in L_{12}.

THEOREM 4.9 (Lee [85]). *A lattice variety $\mathcal{V}$ is neardistributive if and only if $\mathcal{V}$ is semidistributive and contains neither L_{11} or L_{12}.*

PROOF. The forward implication follows immediately from the remarks above. Conversely, suppose that $L_{11}, L_{12} \notin \mathcal{V}$ and $\mathcal{V}$ is semidistributive but not neardistributive. We show that this leads to a contradiction. By duality we may assume that (SD_2^-) does not hold in $\mathcal{V}$, so for some lattice $L \in \mathcal{V}$, $x, y, z, a, c \in L$ we have $x(y + x(z + xy)) = a < c = x(y + z)$. Let $\overline{L}$ be a homomorphic image of L such that $\overline{c}/\overline{a}$ is a critical quotient in $\overline{L}$. Clearly $\overline{L}$ excludes L_{11} and L_{12}, hence Theorem 4.8 implies that $\overline{c}/\overline{a}$ is prime and $\overline{a}$ is meet irreducible. Thus $\overline{a} = \overline{x}$ or $\overline{a} = \overline{y} + \overline{x}(\overline{z} + \overline{xy})$. But $\overline{c} \le \overline{x}$, whence $\overline{a} = \overline{y} + \overline{x}(\overline{z} + \overline{xy}) \ge \overline{y}$. This however is impossible, since $\overline{x} \ge \overline{y}$ implies $\overline{a} = \overline{y} + \overline{x}(\overline{zy}) = \overline{y} + \overline{c} = \overline{c}$. $\square$

For finite lattices we can get an even stronger result.

THEOREM 4.10 (Lee [85]). *A finite lattice L is neardistributive if and only if L is semidistributive and excludes L_{11} and L_{12}.*

PROOF. The forward direction follows immediately from Theorem 4.9. Conversely, suppose L is finite, semidistributive and excludes L_{11} and L_{12}, but is not neardistributive. Then by Theorem 4.9, $\{L\}^{\mathcal{V}}$ contains a lattice K, where K is one of the lattices $M_3, L_1, \ldots, L_5, L_{11}, L_{12}$. Since K is subdirectly irreducible, Jónsson's Lemma implies $K \in \mathbf{HS}\{L\}$. It is also easy to check that every choice of K satisfies Whitman's condition (W), hence Theorem 2.47 implies that K is isomorphic to a sublattice of L. This however contradicts the assumption that L is semidistributive and excludes L_{11}, L_{12}. $\square$

It is not known whether the above theorem also holds for infinite lattices. Note that Theorem 4.8 implies that any finite subdirectly irreducible neardistributive lattice has a unique critical quotient.

Rose [84] observed that any semidistributive lattice which contains a cycle must include either L_{11} or L_{12} (refer to Section 2.3 for the definition of a cycle). This follows easily from Corollary 4.7 and the fact that if $p_1 \sigma p_2 \sigma \ldots \sigma p_n \sigma p_0$ is a cycle then

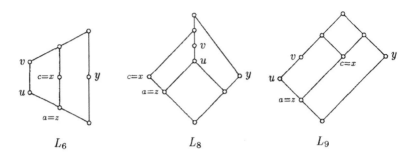

Figure 4.9

$\mathrm{con}(p_i, p_{i_*}) \subseteq \mathrm{con}(p_{i+1}, p_{i+1_*})$ (see Figure 2.6), whence all the quotients p_i/p_{i_*} generate the same congruence. In particular, Theorem 4.10 and Corollary 2.35 therefore imply
that every finite neardistributive lattice is bounded.

Almost distributive lattices. The next definition is also from Lee [85]. A lattice or
a lattice variety is said to be *almost distributive* if it is neardistributive and satisfies the
inclusion

$$(\mathrm{AD}^\cdot) \qquad v(u + c) \le u + c(v + a), \qquad \text{where } a = xy + xz, \; c = x(y + xz),$$

and it dual (AD^+).

Every distributive lattice is almost distributive, since the distributive identity implies
$u + c(v + a) = (u + c)(u + v + a) \ge (u + c)v$. On the other hand L_{11} and L_{12} fail to be
almost distributive, since they are not neardistributive. Further investigation shows that
$(\mathrm{AD}^\cdot)$ fails in L_6, L_8 and L_9 (see Figure 4.9), and by duality (AD^+) does not hold in L_7
and L_{10}, while the next lemma shows that N, L_{13}, L_{14} and L_{15} are almost distributive.

Recall from Section 2.3 Day's construction of "doubling" a quotient u/v in a lattice
L to obtain a new lattice $L[u/v]$. Here we only need the case where L is a distributive
lattice D, and $u = v = d \in D$. In this case we denote the new lattice by $D[d]$. Note that
N, L_{13}, L_{14} and L_{15} can be obtained from a distributive lattice in this way.

LEMMA 4.11 (Lee [85]). *For any distributive lattice D and $d \in D$, the lattice $D[d]$ is
almost distributive.*

PROOF. By duality it suffices to show that $D[d]$ satisfies $(\mathrm{SD}_2^\cdot)$ and $(\mathrm{AD}^\cdot)$. If $(\mathrm{SD}_2^\cdot)$ fails,
then we can find $x, y, z, a, c \in D[d]$ such that $x(y + x(z + xy)) = a < c = x(y + z)$. Let
$\overline{u}$ be the image of $u \in D[d]$ under the natural epimorphism $D[d] \twoheadrightarrow D$ (i.e. $\overline{u} = u$ for
all $u \ne d$, and $\overline{(d, 0)} = \overline{(d, 1)} = d$). Since D is distributive, we must have $\overline{a} = \overline{c}$, whence
$a = (d, 0)$ and $c = (d, 1)$. Clearly a is meet irreducible by the construction of $D[d]$, and
this leads to a contradiction as in Theorem 4.9. To show that $(\mathrm{AD}^\cdot)$ holds in $D[d]$, let us
now denote by u, v, x, y, z, a, c the elements of $D[d]$ corresponding to an assignment of the
(same) variables of $(\mathrm{AD}^\cdot)$. If $a = c$, then $(\mathrm{AD}^\cdot)$ obviously holds. If $a \ne d$ (i.e. $a < c$),
then the distributivity of D again implies that $\overline{a} = \overline{c}$, hence $a = (d, 0)$ and $c = (d, 1)$. Now

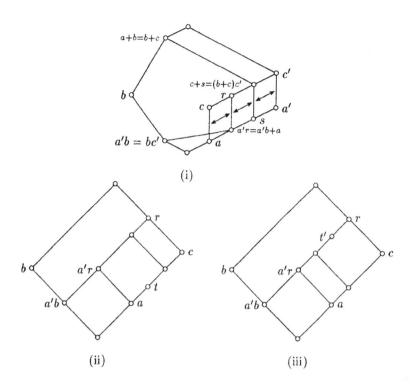

Figure 4.10

$v \leq a$ implies $u + c(v + a) = u + a \geq v \geq v(a + c)$, while $v \not\leq a$ and the meet irreducibility of a imply $v + a \geq c$, whence $u + c(v + a) = u + c \geq v(u + c)$. Thus (AD·) holds in all cases. □

Of course not every almost distributive lattice is of the form $D[d]$ (take for example 2×2, or any one element lattice), but we shall see shortly that all subdirectly irreducible almost distributive lattices can indeed be characterized in this way.

LEMMA 4.12 (Jónsson and Rival [**79**]). *Let L be a semidistributive lattice which excludes L_{12}, and suppose that $a, b, c, a', b' \in L$ with $N(c/a, b)$ and $c/a \nearrow c'/a'$. Set $r = a'b + c$ and $s = (b + c)a'$. Then*

(i) $c/a \nearrow_\beta r/a'r$ *or L includes L_8 or L_{10};*

(ii) $c'/a' \searrow_\beta c + s/s$ *or L includes L_7 or L_9;*

(iii) $r/a'r \nearrow_\beta c + s/s$ *or L includes L_6.*

PROOF. (i) Note that the lattice in Figure 4.10 (i) is isomorphic to F_4 in Figure 4.5. Assume L excludes L_8 and L_{10} (as well as L_{12}), and take $x = c$, $y = a'$, $z = b$ in Lemma 4.5 (iii). Then L is a homomorphic image of F_3, and since we have $N(c'/a', b)$ and

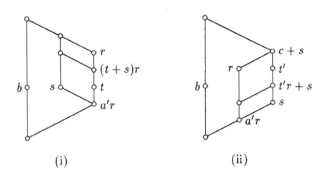

Figure 4.11

$ra' = a'b + a$ in F_3, the same is true in L. For $t \in c/a$ we must have $(t + a'b)c = t$, else
$b, a'r, t, (t + a'b)c$ generate a sublattice isomorphic to L_{10} (see Figure 4.10 (ii)). Similarly,
for $t' \in r/a'r$ we must have $ct' + a'b = t'$ to avoid $L_8(t', c, b)$ (see Figure 4.10 (iii)). This
shows that c/a transposes up onto $r/a'r$ and also proves that this transposition is bijective.
(ii) is dual to (i). Lastly (iii) hold because for $t \in r/a'r$ and $t' \in c + s/s$, we must have
$(t + s)r = t$ and $t'r + s = t'$, to avoid $L_6((t + s)r/t, s, b)$ and $L_6(t'r + s/t', r, b)$ (see
Figure 4.11).   □

Characterizing almost distributive varieties. The next theorem is implicit in Jónsson
and Rival [**79**] and appeared in the present form in Rose [**84**].

THEOREM 4.13 *Let L be a subdirectly irreducible semidistributive lattice. Then the
following conditions are equivalent:*

(i) *L excludes $L_6, L_7, L_8, L_9, L_{10}, L_{11}, L_{12}$;*

(ii) *L has at most one N-quotient;*

(iii) *$L \cong D[d]$ for some distributive lattice D and some $d \in D$.*

PROOF. Assume that (i) holds, and consider an N-quotient u/v in L. By Theorem 4.8, L
has a unique critical quotient which we denote by c/a. It follows that c/a is prime, and
Lemma 1.11 implies that u/v projects weakly onto c/a, say

$$u/v = x_0/y_0 \nearrow_w x_1/y_1 \searrow_w \dots \nearrow_w x_n/y_n = c/a.$$

Of course this implies that $x_i/y_i \in \text{con}(u, v)$ for each $i = 0, 1, \dots, n$, and since u/v
is assumed to be an N-quotient, we have $N(u/v, b)$ for some $b \in L$. Thus Corollary 4.7
implies $N(x_i/y_i, b)$ for each i. In particular, it follows that c/a is an N-quotient. We show
that it is the only one. Note that $x_i/y_i \searrow_w x_{i+1}/y_{i+1}$ implies $x_{i+1}/y_{i+1} \nearrow y_i + x_{i+1}/y_i$,
whence by Lemma 4.12 (i), (ii) and (iii) this transposition is bijective. Similarly the dual
of Lemma 4.12 shows that $x_i/y_i \nearrow_w x_{i+1}/y_{i+1}$ implies $x_{i+1}/y_{i+1} \searrow_\beta x_i/x_iy_{i+1}$. Hence c/a
is projective to a subquotient of u/v (see Figure 4.12). By Theorem 4.8 this subquotient
must equal c/a, otherwise L would have two critical quotients. Furthermore $u < v$ implies

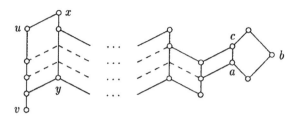

Figure 4.12

that u/c is also an N-quotient, and for the same reason as above, c/a would have to be a subquotient of u/c. But this is clearly impossible, hence $u = c$, and similarly $v = a$.

If (ii) holds then L excludes L_{11} and L_{12}, so again Theorem 4.8 implies that c/a is a prime quotient and $\mathrm{con}(a,c)$ identifies no two distinct elements of L except a and c. Let $D = L/\mathrm{con}(a,c)$. Clearly D cannot include M_3, otherwise the same result would be true for L, contradicting semidistributivity. D also excludes the pentagon N, since $\mathrm{con}(a,c)$ collapses the only N-quotient of L. Hence D is distributive. Let $d = a, c \in D$, then it is easy to check that the map $x \mapsto \{x\}$ $(x \neq c, a), c \mapsto (d,1)$ and $a \mapsto (d,0)$ is an isomorphism from L to $D[d]$. To prove that (iii) implies (i), we first note that since D is distributive, the natural homomorphism from $D[d]$ to D must collapse any N-quotient in $D[d]$. Hence $(d,1)/(d,0)$ is the only N-quotient, and as each of the lattices $L_6, L_7, \ldots, L_{12}$ has at least two N-quotients, (i) must hold. □

The following corollary summarizes the results that have been obtained about almost distributive lattices and varieties.

COROLLARY 4.14

(i) A subdirectly irreducible lattice L is almost distributive if and only if $L \cong D[d]$ for some distributive lattice D and $d \in D$.

(ii) A lattice variety is almost distributive if and only if it is semidistributive and contains none of the lattices $L_6, L_7, \ldots, L_{12}$.

(iii) Every finitely generated subdirectly irreducible almost distributive lattice is finite.

(iv) Every almost distributive variety that has finitely many subvarieties is generated by a finite lattice.

(v) Every join irreducible almost distributive variety of finite height is generated by a finite subdirectly irreducible lattice.

(vi) Every finite almost distributive lattice is bounded (in the sense of Section 2.3).

PROOF. (i) The forward implication follows from Theorem 4.13, since L is certainly semidistributive and, as $L_6, L_7, \ldots, L_{12}$ all fail to be almost distributive, L must exclude these lattices. Theorem 4.11 provides the reverse implication.

(ii) The forward direction is trivial. Conversely, if $\mathcal{V}$ is semidistributive and contains none of $L_6, L_7, \ldots, L_{12}$, then Theorem 4.11 and Theorem 4.13 imply that every subdirectly irreducible member, and hence every member of $\mathcal{V}$ is almost distributive.

(iii) If the lattice $L \cong D[d]$ in (i) is finitely generated, so is the distributive lattice D. It follows that D is finite, and since $|L| = |D[d]| = |D| + 1$, L is also finite. Now (iv) and (v) follow from Lemma 2.7, and (vi) is a consequence of Lemmas 2.40 and 2.41. $\square$

In particular the last result shows that any finite subdirectly irreducible almost distributive lattice is a splitting lattice.

Part (iii) above says that almost distributive varieties are locally finite, and this is the reason why they are much easier to describe than arbitrary varieties. More generally we have the following result:

LEMMA 4.15 (Rose [84]). *Let L be a finitely generated subdirectly irreducible lattice all of whose critical quotients are prime. If c/a is a critical quotient of L, and if $L/\mathrm{con}(a, c)$ belongs to a variety that is generated by a finite lattice, then L is finite.*

PROOF. Since every critical quotient of L is prime, each congruence class of $\mathrm{con}(a, c)$ has at most two elements. Thus the assumption that L is infinite implies that $L/\mathrm{con}(a, c)$ is infinite as well. However, this leads to a contradiction, since the lattice $L/\mathrm{con}(a, c)$ is finitely generated and belongs to a variety generated by a finite lattice, hence $L/\mathrm{con}(a, c)$ must be finite. $\square$

Subdirectly irreducible lattices of the form $D[d]$. We continue our investigation of semidistributive lattices that exclude L_{11} and L_{12}, which will then lead to a characterization of all the finite subdirectly irreducible almost distributive lattices.

LEMMA 4.16 (Rose [84]). *Let L be a subdirectly irreducible semidistributive lattice that excludes L_{11} and L_{12}, and suppose c/a is the (unique) critical quotient of L. Then*

(i) *the sublattices $[a)$ and $(c]$ of L are distributive;*

(ii) *for any nontrivial quotient u/v of $(c]$ there exist $b, v' \in L$ with $v \le v' < u$ such that $N(c/a, b)$, $b \le u$, $b \not\le v'$ and $v' + b/v' \searrow a + b/(a + b)v'$ (Figure 4.13).*

(iii) *if $u \succ v = c$ in (ii), then we also have $u = a + b$.*

PROOF. (i) By semidistributivity, L excludes M_3. Suppose that for some $u, v, b \in L$ we have $N(u/v, b)$. Then Corollary 4.7 implies $N(c/a, b)$, whence $b \notin [a)$ and $b \notin (c]$. It follows that $[a)$ and $(c]$ also exclude the pentagon, and are therefore distributive.

(ii) Choose a shortest possible sequence

$$u/v = x_0/y_0 \sim_w x_1/y_1 \sim_w \ldots \sim_w x_n/y_n = c/a.$$

Since $v \ge c$, we must have $n \ge 3$. Suppose that $u/v \nearrow_w x_1/y_1 \searrow_w x_2/y_2$. By the minimality of n, only part (i) or (ii) of Lemma 4.3 can apply. That is, there exist $a', b, c', u' \in L$ with $N(c'/a', b)$ and $v < u' \le u$ such that $b/bc' \subseteq u/v$ or $u'/v \nearrow_w a' + b/b$. But Corollary 4.7 implies $N(c/a, b)$, which is impossible since in both cases $b \ge a$. Thus we must have $u/v \searrow_w x_1/y_1 \nearrow x_2/y_2$. By the dual of Lemma 4.3 (i) and (ii), there exist $a', b, c', v' \in L$ with $N(c'/a', b)$ and $v \le v' < u$ such that either $a' + b/b \subseteq u/v$ or $u/v' \searrow_w b/bc'$. Only the latter is possible, since we again have $N(c/a, b)$ by Corollary 4.7. Now $a, b < u$ implies

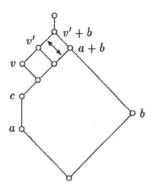

Figure 4.13

$a + b \leq u$, and $a < v'$ implies $v' + b = v' + (a + b)$, hence $v' + b/v' \searrow a + b/(a + b)v'$.
Also $bv' = bc' \neq b$ implies $b \not\leq v$, and the bijectivity of the transposition follows from the
distributivity of $[a)$ (part (i)). (iii) is a special case if (ii).   □

THEOREM 4.17 (Rose [84]). *Let D be a finite distributive lattice and $d \in D$. Then $D[d]$
is subdirectly irreducible if and only if all of the following conditions hold:*

(i) *every cover of d is join reducible,*

(ii) *every dual cover of d is meet reducible, and*

(iii) *every prime quotient in D is projective to a prime quotient p/q with $p = d$ or $q = d$.*

PROOF. Suppose $L = D[d]$ is subdirectly irreducible. Let $a = (d, 0)$ and $c = (d, 1)$. Notice
that $D = d$ implies that (i), (ii) and (iii) are satisfied vacuously. If $u \in D$ covers d, then
u covers c in L, whence by Lemma 4.16 (iii) there exists $b \in L$ noncomparable with c,
and $u = a + b$. Thus u is join reducible in L and also in D. Dually, every element that is
covered by d is meet irreducible. To prove (iii), consider a prime quotient $u/v \neq c/a$ in L,
and choose a sequence

$$u/v = x_0/y_0 \sim_w x_1/y_1 \sim_w \ldots \sim_w x_n/y_n \supseteq c/a$$

with n as small as possible. For $i < n$ none of the quotients x_i/y_i contains c/a, and is
therefore isomorphic to $\overline{x_i}/\overline{y_i}$ in D (where $\overline{x}$ denotes the image of x under the natural
epimorphism $D[d] \to D$). Since D is distributive and $\overline{u}/\overline{v}$ is prime, each $\overline{x_i}/\overline{y_i}$ is prime,
whence $\overline{x_0}/\overline{y_0} \sim \overline{x_1}/\overline{y_1} \sim \ldots \sim \overline{x_n}/\overline{y_n}$. It follows that $\overline{x_n} = d$ or $\overline{y_n} = d$, so (iii) holds with
$p/q = \overline{x_n}/\overline{y_n}$.

Conversely, suppose (i), (ii) and (iii) hold. Since D and hence L are finite, it suffices to
show that every prime quotient of L projects weakly onto c/a. We begin by showing that
every prime quotient $u/v \neq c/a$ is projective to a prime quotient x/y with $x = a$ or $y = c$.
Since c is the only cover of a (and dually), we cannot have $v = a$ or $u = c$. Also if $u = a$

or $v = c$, then we take $x/y = u/v$. Otherwise, using (iii), we may assume by duality, that $\overline{u}/\overline{v}$ is projective to a prime quotient $\overline{x}/d$ with $x \succ c$. Since D is distributive, this means that $\overline{u} \not\leq d$ and $\overline{u}/\overline{v} \nearrow \overline{u} + \overline{x}/\overline{v} + \overline{c} \searrow \overline{x}/\overline{c}$, i.e. $\overline{u} + \overline{c} = \overline{u} + \overline{x} = \overline{x} + \overline{v}$, $\overline{u}(\overline{v} + \overline{c}) = \overline{v}$ and $\overline{x}(\overline{v} + \overline{c}) = \overline{c}$. Hence $u + c = u + x = x + v$, and further more $\overline{v} \neq d$ implies $u(v + c) = v$ and $x(v + c) = c$ (since $a < c$). Thus $u/v \nearrow u + x/v + x \searrow x/c$. Now we apply (i) to obtain $b \in L$ with $\overline{b} < \overline{x}$ and $\overline{x} = \overline{c} + \overline{b}$. It follows that $x = c + b = a + b$ (since a is meet irreducible), while the join-irreducibility of c implies $cb = ab$. Thus we have $N(c/a, b)$, and since $\text{con}(u, v)$ identifies x and c, it also identifies c and a. Consequently L is subdirectly irreducible. $\hspace{2cm} \square$

Varieties covering that smallest nonmodular variety. From the results obtained so far one can now prove the following:

THEOREM 4.18 *The variety $\mathcal{N}$ is covered by precisely three almost distributive varieties, $\mathcal{L}_{13}$, $\mathcal{L}_{14}$ and $\mathcal{L}_{15}$.*

PROOF. With the help of Jónsson's Lemma it is not difficult to check that each of the varieties $\mathcal{L}_i (= \{L_i\}^\mathcal{V})$ cover $\mathcal{N}$ ($i = 1, \ldots, 15$). So let $\mathcal{V}$ be an almost distributive variety that properly includes $\mathcal{N}$. We have to show that $\mathcal{V}$ includes at least one of $\mathcal{L}_{13}$, $\mathcal{L}_{14}$ or $\mathcal{L}_{15}$. Every variety is determined by its finitely generated subdirectly irreducible members, hence we can find such a lattice $L \in \mathcal{V}$ not isomorphic to N or **2**. By Corollary 4.14 (i) and (iii), $L \cong D[d]$ for some finite distributive lattice D, $d \in D$. We show that $D[d]$ contains one of L_{13}, L_{14} or L_{15} as a sublattice. D is nontrivial since $D[d] \not\cong \mathbf{2}$. Let 0_D and 1_D be the smallest and largest element of D respectively. Theorem 4.17 (i) and (ii) imply that $d \neq 0_D, 1_D$. Also, $0_D \prec d \prec 1_D$ would imply $D[d] \cong N$, so by duality we can find $u, v \in D$ such that $v \prec u \prec d$. By Theorem 4.17 (iii) u/v is projective to a prime quotient p/q such that $p = d$ or $q = d$. Since D is distributive and $u < d$, the case $q = d$ is excluded, hence u/v is projective to d/q. Again, by the distributivity of D, $u \neq q$ and therefore $d = u + q$ (see Figure 4.14 (i)). By Theorem 4.17 (ii) u and q are meet reducible, so there exist $x, y \in D$ such that $u = xd$, $q = yd$ and since D is finite we may assume that $x \succ u$ and $y \succ v$. The sublattice D' of D generated by x, d, y is a homomorphic image of the lattice in Figure 4.14 (ii) (the distributive lattice with generators x, d, y and defining relation $d = xd = yd$). Since $x \succ u$ and $y \succ v$, D' must in fact be isomorphic to the lattice in Figure 4.14 (iii) or (iv). Consequently $D'[d]$, as a sublattice of $D[d]$, is isomorphic to L_{14} or L_{15}. A sublattice isomorphic to L_{13} is obtained from the dual case when $d \prec v \prec u$. $\hspace{1cm} \square$

The above theorem and Corollary 4.14 (ii) now imply:

THEOREM 4.19 (Jónsson and Rival [**79**]). *In the lattice Λ of all lattice subvarieties, the variety $\mathcal{N}$ is covered by exactly 16 varieties: $\mathcal{M}_3 + \mathcal{N}$, $\mathcal{L}_1$, $\mathcal{L}_2, \ldots, \mathcal{L}_{15}$.*

Representing finite almost distributive lattices. Building on Theorem 4.17, Lee [**85**] gives another criterion for the subdirect irreducibility of a lattice $D[d]$, where D is distributive, and he also sets up a correspondence between finite subdirectly irreducible almost distributive lattices and certain matrices of zeros and ones. Before discussing his results, we recall some facts about distributive lattices which can be found in [**GLT**].

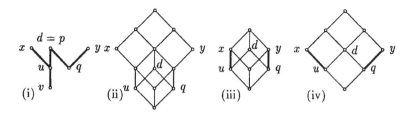

Figure 4.14

LEMMA 4.20 *Let D and D' be finite distributive lattices and denote by $J(D)$ the poset of all nonzero join irreducible elements of D. Then*

(i) *any poset isomorphism from $J(D)$ to $J(D')$ can be extended to an isomorphism from D to D'.*

(ii) *every maximal chain of D has length $|J(D)|$.*

Given a finite distributive lattice D and $d \in D$, let $B = \{b_1, \ldots, b_m\}$ be the set of all meet reducible dual covers of d, $C = \{c_1, \ldots, c_n\}$ the set of all join reducible covers of d, and consider the set

$$X_d(D) = \{x \in D : xd \prec d \prec x + d\}.$$

We define two partitions $\{B_1, \ldots, B_m\}$ and $\{C_1, \ldots, C_n\}$ of $X_d(D)$, referred to as the *natural partitions* of $X_d(D)$, as follows:

$$B_i = \{x \in X_d(D) : xd = b_i\}, \qquad C_j = \{x \in X_d(D) : x + d = c_j\}.$$

(We assume here, and subsequently, that the index i ranges from 1 to m and j ranges from 1 to n.) By the distributivity of D, two blocks B_i and C_j have at most one element in common, so we can define an $m \times n$ matrix $A(D[d])$ of 0's and 1's by $a_{ij} = |B \cap C|$. (If any, and hence all, of the sets B, C or $X_d(D)$ is empty, then $A(D[d]) = ()$, the 0×0 matrix with no entries.)

$A(D[d])$ is called the matrix associated with $D[d]$, but notice that because the elements of B and C were labeled arbitrarily, $A(D[d])$ is determined only up to the interchanging of any rows or any columns. Observe also that $A(D[d])$ does not have any rows or columns with just zeros, since $\{B_i\}$ and $\{C_j\}$ are partitions of the same set $X_d(D)$. As examples we note that $A(\mathbf{2}) = ()$, $A(N) = (1)$, $A(L_{13}) = (1,1)$, $A(L_{14}) = \begin{pmatrix} 1 \\ 1 \end{pmatrix}$ and $A(L_{15}) = \begin{pmatrix} 1 & 0 \\ 0 & 1 \end{pmatrix}$ or $\begin{pmatrix} 0 & 1 \\ 1 & 0 \end{pmatrix}$ (see Figure 4.15).

We will also be concerned with the sublattice D^* of D generated by the set $U = X_d(D) \cup \{d\}$. Let $1^* = \sum U$, $0^* = \prod U$, then clearly the elements $c_1, \ldots, c_n$ will be atoms in the quotient $1^*/d$ and $\sum_j c_j = 1^*$, so by Lemma 1.12, $1^*/d$ is isomorphic to the Boolean algebra $\mathbf{2}^n$, and the elements $c_1, \ldots, c_n$ are the only covers of d in D^*. Dually we have

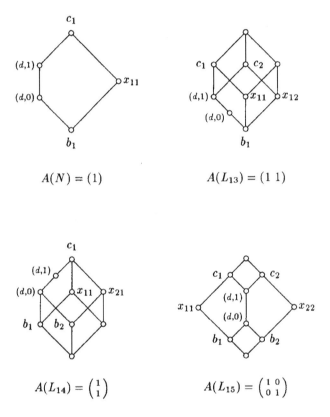

Figure 4.15

that $d/0^*$ is isomorphic to the Boolean algebra 2^m and that $b_1, \ldots, b_m$ are the only dual covers of d in D^*. It follows that D^* has length $m + n$ and therefore, by Lemma 4.20 (ii), $|J(D^*)| = m + n$. We can in fact describe the elements of $J(D^*)$:

LEMMA 4.21 (Lee [85]). $J(D^*) = \{b'_1, \ldots, b'_m, c'_1, \ldots, c'_n\}$ where b'_i is the complement of b_i in $d/0^*$, and $c'_j = \prod C_j$. All elements of $J(D^*)$ are incomparable except: $b'_i \leq c'_j$ if and only if $B_i \cap C_j = \emptyset$.

PROOF. It is clear that the b'_i are distinct atoms of D^* and therefore pairwise incomparable and join irreducible in D^*. As for the c'_j, we first note that in a distributive lattice every join irreducible element is a meet of generators. Thus if c is join irreducible and $c < c'_j$, then $c \leq c'_j x$ for some $x \in U - C_j$. Now $c'_j + d = c_j \neq x + d$, hence $c'_j x d = (c'_j + d)(x + d) = d$, which shows that $c \leq c'_j x \leq d$. But $c'_j \not\leq d$ and therefore c'_j cannot be a join of join irreducibles strictly less that itself. It follows that c'_j is join irreducible. Also $c'_1, \ldots, c'_n$ are pairwise incomparable, since $c'_j \leq c'_k$ for some $j \neq k$ implies $c'_j = c'_j c'_k \leq c'_j x$ for any $x \in C_k$, and $c'_j x \leq d$ as above, which contradicts $c'_j \not\leq d$. Clearly also $c'_j \not\leq b'_i$ for any i and j, since $b'_i \leq d$. Therefore it remains to show that $b'_i \leq c'_j$ if and only if $B_i \cap C_j = \emptyset$. If $x \in B_i \cap C_j$, then $b'_i \leq d$ and $c'_j \leq x$, so $b'_i c'_j = b'_i (dx) c'_j = (b'_i b_i) c'_j = 0^* c'_j = 0^*$ and hence $b'_i \not\leq c'_j$. Conversely $B_i \cap C_j = \emptyset$ implies $C_j \subseteq U - B_i$, and since b'_i is a meet of generators, $b'_i = \prod U - B_i \leq \prod C_j = c'_j$. $\square$

So, given any $m \times n$ matrix $A = (a_{ij})$ of 0's and 1's, we define a finite distributive lattice D_A and an element d_A as follows:

Suppose 2^{m+n} be the Boolean algebra generated by the $m + n$ atoms $p_1, \ldots, p_m$, $q_1, \ldots, q_n$. Put

$$d_A = \sum p_i, \qquad \overline{b}_k = \sum_{i \neq k} p_i \qquad \text{and} \qquad x_{ij} = \overline{b}_i + q_j,$$

and let $X_A = \{x_{ij} : a_{ij} = 1\}$. Now we let D_A be the sublattice of 2^{m+n} generated by $X_A \cup \{d_A\}$.

LEMMA 4.22 (Lee [85]). For no proper subset U of $X_d(D)$ does $U \cup \{d\}$ generate D^*.

PROOF. We may assume that $X_d(D)$ is nonempty. Suppose to the contrary that $U = X_d(D) - \{x_0\}$ for some $x_0 \in X_d(D)$, and $U \cup \{d\}$ generates D^*. Then $x_0 \in C_j$ for some block C_j of the natural partition $\{C_1, \ldots, C_n\}$ of $X_d(D)$. Let $c'_j = \prod C_j \in D^*$. By Lemma 4.21 c'_j is join irreducible, and since $U \cup \{d\}$ is a generating set, c'_j is the meet of a subset V of $U \cup \{d\}$. Notice that $x \prec x + d = c'_j$ for each $x \in C_j$, so by the dual of Lemma 1.12 C_j generates a Boolean algebra with least element c'_j. Hence V is not a proper subset of C_j and, as $x_0 \notin V$, we also cannot have $V = C_j$. Choose $x \in V - C_j$. Then $x + d \neq x_0 + d = c_j$, so

$$d = (x + d)(x_0 + d) = x x_0 + d \geq x x_0 \geq \prod C_j = \prod V = c'_j.$$

However this contradicts $c'_j \not\leq d$. $\square$

THEOREM 4.23 (Lee [85]). Let D be a finite distributive lattice and $d \in D$. Then the following are equivalent:

(i) $D[d] \cong D_A[d_A]$ where $A = A(D[d])$ and $d \in D$ corresponds to $d_A \in D_A$;

(ii) *the set $X_d(D) \cup \{d\}$ generates D (i.e. $D^* = D$);*

(iii) *$D[d]$ is subdirectly irreducible.*

PROOF. (i) implies (ii): Let $d_A, \overline{b_k}, x_{ij}$ and X_A be defined as above. Clearly $x_{ij} d_A \prec d_A \prec x_{ij} + d_A$ for all i, j, hence $X_A \subseteq X_{d_A}(D_A)$. Since $X_A \cup \{d\}$ generates D_A, so does $X_{d_A}(D_A) \cup \{d\}$. Notice that $D_A^* = D_A$, and by Lemma 4.22 $X_A = X_{d_A}(D_A)$.

(ii) implies (i): Again suppose 2^{m+n} is the Boolean algebra generated by the atoms $p_1, \ldots, p_m, q_1, \ldots, q_n$, and let $\overline{c_j} = d_A + q_j$. We claim that the elements $\overline{b_1}, \ldots, \overline{b_m}, \overline{c_1}, \ldots, \overline{c_n}$ are all in D_A. This follows because $x_{ij} d_A = (\overline{b_i} + q_j) d_A = \overline{b_i} d_A + q_j d_A = \overline{b_i}$ and $x_{ij} + d_A = \overline{b_i} + q_j + d_A = d_A + q_j = \overline{c_j}$ for all i, j, and $A = A(D[d])$ has no rows or columns of zeros, hence for any given i (or j) there exists j (respectively i) such that $x_{ij} \in X_A$. Clearly the $\overline{c_j}$ are covers of d_A, and they are the only ones, since by Lemma 1.12 $\sum \overline{c_j} = \sum X_A = 1_A^*$. Dually the $\overline{b_i}$ are all the dual covers of d_A. Let $\{B_1, \ldots, B_m\}$ and $\{C_1, \ldots, C_n\}$ be the natural partitions of $X_{d_A}(D_A) = X_A$. By Lemma 4.21 $J(D_A^*) = \{\overline{b_1'}, \ldots, \overline{b_m'}, \overline{c_1'}, \ldots, \overline{c_n'}\}$ where $\overline{b_i'} = p_i$ and $\overline{c_j'} = \prod \overline{C_j}$. Now $B_i \cap C_j \neq \emptyset$ iff $a_{ij} = 1$ in $A(D[d])$ iff $\overline{b_i} \leq x_{ij} \leq \overline{c_j}$ in D_A iff $\overline{B_i} \cap \overline{C_j} \neq \emptyset$. Consequently the map $b_i' \mapsto \overline{b_i'}$, $c_j' \mapsto \overline{c_j'}$ from $J(D^*)$ to $J(D_A^*)$ is a poset isomorphism which extends to an isomorphism $D^* \cong D_A^*$ by Lemma 4.20 (i). D^* is the sublattice of D generated by $X_d(D) \cup \{d\}$, so by assumption $D^* = D$, and we always have $D_A^* = D_A$. Clearly also $d = \sum b_i'$ is mapped to $d_a = \sum \overline{b_i'}$ by the isomorphism.

(ii) implies (iii): We verify that the conditions (i), (ii) and (iii) of Theorem 4.17 hold. By Lemma 1.12, the join reducible covers of d in $D^* = D$ are in fact all the covers of d, and dually, which implies that the first two conditions hold. Also, if $u \prec v$ in D, then the length of $D/\mathrm{con}(u,v)$ is less that the length of D. It follows that $\mathrm{con}(u,v)$ identifies d with one of its covers or dual covers, hence condition (iii) of Theorem 4.17 holds.

(iii) implies (ii): Suppose $D[d]$ is subdirectly irreducible, but D^* is a proper sublattice of D. Let 0^* be the smallest and 1^* the largest element of D^*, and choose an element $z \in D - D^*$.

Case 1: $z \not\leq 1^*$ or $z \not\geq 0^*$. Then one of the quotients $z + 1^*/1^*$ or $0^*/z0^*$ is nontrivial. Observe that in any distributive lattice, if $v < u \leq v' < u'$, then the quotients u/v and u'/v' cannot project onto each other. Hence no prime quotients in $z + 1^*/1^*$ or $0^*/z0^*$ project onto any prime quotient p/q with $p = d$ or $q = d$, since $p, q \in D^*$ by condition (i) and (ii) of Theorem 4.17. This however contradicts condition (iii) of the same theorem.

Case 2: $0^* \leq z \leq 1^*$. Choose z such that the height of z is as large as possible, and let z^* be a cover of z. Then $z^* \in D^*$, and z^* is the only cover of z, else z would be the meet of two elements from D^* and would also belong to D^*. By Theorem 4.17 (iii), the prime quotient z^*/z projects onto a prime quotient p/q with $p = d$ or $q = d$, and since D is distributive, this implies $z^*/z \nearrow u/v \searrow p/q$ for some quotient u/v. Since z^* is the unique cover of z, we must have $u/v = z^*/z \searrow p/q$. Suppose $p = d$. Then $z^*/z \searrow d/zd$, and the two quotients are distinct, otherwise Theorem 4.17 (ii) implies $z \in D^*$. Consequently $z^*/d \searrow_\beta z/zd$. As before Lemma 1.12 implies that $1^*/d$ is a Boolean algebra, hence z^*/d is a Boolean algebra, and so is z/zd (via the bijective transposition). Therefore z is the join of the atoms of z/zd, which are in fact elements of $X_d(D)$. This implies $z \in D^*$, a contradiction. Next suppose $q = d$. Since, by Lemma 1.12, $1^* \subseteq D^*$ we would again have $z \in D^*$, a contradiction. Thus we conclude that $D^* = D$. □

Given any matrix A of 0's and 1's with no rows or columns of zeros, the equivalence of

(ii) and (iii) tells us that $D_A[d_A]$ is subdirectly irreducible. Conversely, for any subdirectly irreducible lattice $D[d]$, the matrix $A(D[d])$ has no rows or columns of zeros, and it is not difficult to see that, up to the interchanging of some rows or columns, the matrices A and $A(D_A[d_A])$ are the same. Furthermore, given any lattice $D[d]$ and $X' \subseteq X_d(D)$, the sublattice D' generated by $X' \cup \{d\}$ is subdirectly irreducible, and by Lemma 4.22 $X_d(D') = X'$. Rephrased in terms of the matrices that represent the lattices D and D' we have the following:

COROLLARY 4.24 *Let* $A = A(D[d])$ *for some finite distributive lattice* D, $d \in D$, *and suppose* D' *is the sublattice generated by some* $X' \subseteq X_d(D)$. *Then the matrix* A' *which represents* $D'[d]$ *is obtained from* A *by changing each 1 corresponding to an element of* $X_d(D) - X'$ *to 0 and deleting any rows or columns of zeros that may have arisen. Conversely any matrix obtained from* A *in this way represents a (subdirectly irreducible) sublattice of* $D[d]$.

Covering chains of almost distributive varieties. The next lemma, which was proved by Rose [84] directly from Theorem 4.17, can now be derived from the above corollary.

LEMMA 4.25 *Let* L *be a finite subdirectly irreducible almost distributive lattice,* $L \not\cong \mathbf{2}, N$.

(i) *If* $L_{14}, L_{15} \notin \{L\}^V$ *then* $L \cong L_{13}^k$,

(ii) *if* $L_{13}, L_{15} \notin \{L\}^V$ *then* $L \cong L_{14}^k$ *and*

(iii) *if* $L_{13}, L_{14} \notin \{L\}^V$ *then* $L \cong L_{15}^k$ *for some* $k \in \omega$ *(see Figure 2.2).*

PROOF. (i) By Corollary 4.14 (i) $L \cong D[d]$ for some finite distributive lattice D and $d \in D$. Let $A = A(D[d])$ be the matrix representing $D[d]$ and suppose A has more than one row. If A has no column with two 1's in it, then it has at least two columns (since it has at least two rows, and no rows of 0's), and we can therefore find two entries equal to 1 in two different columns and rows. Deleting all other rows and columns, it follows from Corollary 4.24 that L_{15} is a sublattice of L. Hence if $L_{14}, L_{15} \notin \{L\}^V$, then A has only one row with all entries equal to 1. This is the matrix representing L_{13}^k, where $k + 2$ is the number of columns of A (see Figure 4.16 (i)). Similar arguments prove (ii) and (iii).□

THEOREM 4.26 (Rose [84]). *For each* $i \in \{13, 14, 15\}$ *and* $n \in \omega$ *the variety* $\mathcal{L}_i^{n+1}$ *is the only join irreducible cover of* $\mathcal{L}_i^n$.

PROOF. Let $i = 13$ and suppose $\mathcal{V}$ is a join irreducible variety that covers $\mathcal{L}_{13}^n = \{L_{13}^n\}^V$. $\mathcal{V}$ must be almost distributive, otherwise, by Corollary 4.14 (ii), $\mathcal{V}$ contains one, say L, of the lattices $M_3, L_1, L_2, \ldots, L_{12}$, in which case $\mathcal{V} \geq \mathcal{L}_{13}^n + \{L\}^V > \mathcal{L}_{13}^n$, hence either $\mathcal{V}$ is not a cover of $\mathcal{L}_{13}^n$ or $\mathcal{V}$ is join reducible. $\mathcal{V}$ is of finite height, thus by Corollary 4.14 (i), (iii) and Lemma 2.7 $\mathcal{V}$ is generated by a finite subdirectly irreducible lattice $L = D[d]$, where D is distributive. Since $\mathcal{V}$ is join irreducible, $L_{14}, L_{15} \notin \mathcal{V}$ so by Lemma 4.25 (i) $L = L_{13}^k$, and since $\mathcal{V}$ covers $\mathcal{L}_{13}^n$, we must have $k = n + 1$. The proof for $i = 14$ and 15 is completely analogous. □

The smallest subdirectly irreducible almost distributive lattice that is not of the form $\mathbf{2}$, N or L_i^n for $i = 13, 14, 15$, $n \in \omega$ is represented by the matrix $\begin{pmatrix} 1 & 1 \\ 1 & 0 \end{pmatrix}$ (see Figure 4.16 (iii)).

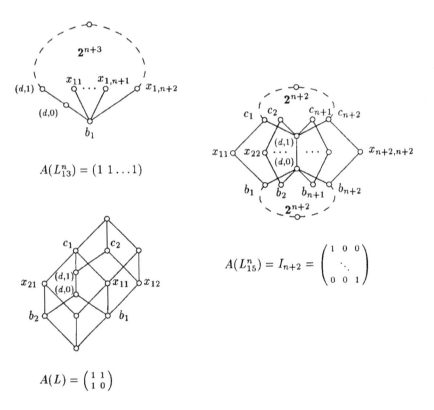

Figure 4.16

Further results on almost distributive varieties.

THEOREM 4.27 (Lee [85]). *Every almost distributive lattice variety of finite height has only finitely many covers.*

PROOF. Let $\mathcal{V}$ be an almost distributive variety of finite height. Then $\mathcal{V}$ is generated by finitely many subdirectly irreducible almost distributive lattices, and by Corollary 4.14 (i) these lattices are of the form $D_1[d_1], \ldots, D_n[d_n]$ for some finite distributive lattices $D_1, \ldots, D_n$. Let $k = \max\{|X_{d_i}(D_i)| : i = 1, \ldots, n\}$. By Corollary 4.14 (iv), each join irreducible cover of $\mathcal{V}$ is generated by a finite subdirectly irreducible lattice $D[d]$, and clearly we must have $|X_d(D)| = k + 1$. By Theorem 4.23 $D[d]$ can be represented by a matrix of 0's and 1's with at most $k + 1$ rows and $k + 1$ columns, hence $\mathcal{V}$ has only finitely many join irreducible covers. On the other hand, each join reducible cover of $\mathcal{V}$ is a join of $\mathcal{V}$ and a join irreducible cover of a subvariety of $\mathcal{V}$. Therefore $\mathcal{V}$ also has only finitely many join reducible covers. □

LEMMA 4.28 (Lee [85]). *Let $\overline{D}$ be a sublattice of a finite distributive lattice D, and $d \in \overline{D}$. If $\overline{D}[d]$ is subdirectly irreducible, then $\overline{D}[d] \cong D'[d]$, where D' is generated by d and a subset of $X_d(D)$.*

PROOF. Let $\{B_1, \ldots, B_m\}$ and $\{C_1, \ldots, C_n\}$ be the natural partitions of $X_d(\overline{D})$, and let $b_i = xd$ with $x \in B_i$, $c_j = x + d$ with $x \in C_j$. Choose $b'_1, \ldots, b'_m, c'_1, \ldots, c'_n \in D$ such that $b_i \leq b'_i \prec d$ and $d \prec c'_j \leq c_j$. For each $x \in X_d(\overline{D})$ we have $x \in B_i \cap C_j$ for unique i, j, in which case we define $x' = xc'_j + b'_i$. By distributivity $x'd = (xc'_j + b'_i)d = xd + b'_i$ and $x' + d = (x + d)c'_j$, hence $xd = b_i$ implies $x'd = b'_i$, and $x + d = c_j$ implies $x' + d = c'_j$. It follows that the set $X' = \{x' : x \in X_d(\overline{D})\}$ is a subset of $X_d(D)$, and since the elements b'_i and c'_j all have to be distinct, the map $x \mapsto x'$ is bijective. Let D' be the sublattice of D generated by $X' \cup \{d\}$. By Lemma 4.22 $X_d(D') = x'$. We show that $\overline{D}$ and D' have the same matrix representation, then it follows from Theorem 4.23 that $\overline{D} \cong D'$. By Lemma 1.12 the elements $c'_1, \ldots, c'_n$ are all the (join reducible) covers of d, and dually for $b'_1, \ldots, b'_m$. Let $B'_i = \{x' \in X' : x'd = b'_i\}$ and $C'_j = \{x' \in X' : x' + d = c'_j\}$ be the blocks of the natural partitions of X'. Clearly $x \in B_i$ implies $x' \in B'_i$ for all i, and the converse must also hold, since the map $x \mapsto x'$ is bijective and the blocks B_i, b'_i are finite. Similarly $x \in C_j$ if and only if $x' \in C'_j$. Hence $|B_i \cap C_j| = |B'_i \cap C'_j|$, which implies $A(\overline{D}[d]) = A(d'[d])$. □

LEMMA 4.29 (Lee [85]). *Let D be a finite distributive lattice, $d \in D$, and let D^* be the sublattice of D generated by $X_d(D) \cup \{d\}$. Then $D^*[d]$ is a retract of $D[d]$. In particular, $D^*[d]$ is the smallest homomorphic image of $D[d]$ separating $(d, 0)$ and $(d, 1)$.*

PROOF. Since $(d, 0) \prec (d, 1)$, by Lemma 1.11 there is a unique subdirectly irreducible homomorphic image $\overline{D[d]}$ of $D[d]$ such that $\overline{(d, 1)/(d, 0)}$ is a critical quotient of $\overline{D[d]}$. By Theorem 4.23 $D^*[d]$ is also subdirectly irreducible with critical quotient $(d, 1)/(d, 0)$, hence $D^*[d]$ is isomorphic to its image $\overline{D^*[d]} \subseteq \overline{D[d]}$. We have to show that $\overline{D^*[d]} = \overline{D[d]}$. The epimorphism $D[d] \twoheadrightarrow \overline{D[d]}$ induces an epimorphism $D \twoheadrightarrow \overline{D}$, where $\overline{D}$ is obtained from $\overline{D[d]}$ by collapsing the quotient $\overline{(d, 1)/(d, 0)}$. Then $D^* \cong \overline{D^*} \subseteq \overline{D}$, and it suffices to show that $\overline{D^*} = \overline{D}$. Consider $x \in D$ such that $\overline{x} \in X_{\overline{d}}(\overline{D})$. x must be noncomparable with d, so we can find $b, c \in D$ with $xd \leq b \prec d \prec c \leq x + d$. Let $x_0 = xc + b$, then one easily

$$A(K_i) = \begin{pmatrix} 1 & 1 & 0 & 0 & \cdots & 0 & 0 \\ 0 & 1 & 1 & 0 & \cdots & 0 & 0 \\ 0 & 0 & 1 & 1 & \cdots & 0 & 0 \\ \vdots & \vdots & \vdots & \vdots & \ddots & \vdots & \vdots \\ 0 & 0 & 0 & 0 & \cdots & 1 & 1 \\ 1 & 0 & 0 & 0 & \cdots & 0 & 1 \end{pmatrix}$$

Figure 4.17

checks that $x_0 \in X_d(D^*)$ and $b, c \in D^*$. Notice that $\overline{d} \neq \overline{c}$, for otherwise the epimorphism $D[d] \twoheadrightarrow \overline{D[d]}$ would collapse the quotient $c/(d,1)$ and, as x_0, $(d,0)$ and $(d,1)$ generate a pentagon, it would also identify $(d,0)$ and $(d,1)$. Similarly $\overline{b} \neq \overline{d}$, whence $\overline{b} \prec \overline{d} \prec \overline{c}$ in $\overline{D^*}$. Because $\overline{x} \in X_{\overline{d}}(\overline{D})$, it follows that $\overline{x}\overline{d} = \overline{b}$, $\overline{x} + \overline{d} = \overline{c}$ and since $\overline{b}, \overline{c} \in \overline{D^*}$, we in fact have $\overline{x} \in X_{\overline{d}}(\overline{D^*})$. Thus $X_{\overline{d}}(\overline{D}) \subseteq X_{\overline{d}}(\overline{D^*}) \subseteq \overline{D^*} \subseteq \overline{D}$. By Theorem 4.23 $X_{\overline{d}}(\overline{D}) \cup \{d\}$ is a generating set for $\overline{D}$, hence $\overline{D}^* = \overline{D}$, and therefore $D^*[d] \cong \overline{D^*[d]} = \overline{D[d]}$. □

LEMMA 4.30 (Lee [85]). *Let D be a finite distributive lattice and $d \in D$. Then every subdirectly irreducible member of $\{D[d]\}^V$ is isomorphic to $D'[d]$, where D' is a sublattice of D generated by d and a subset of $X_d(D)$.*

PROOF. Let L be a subdirectly irreducible member of $\{D[d]\}^V$. By Jónsson's Lemma $L \in \mathbf{HS}\{D[d]\}$, so there is a sublattice L_0 of $D[d]$ and an epimorphism $f : L_0 \twoheadrightarrow L$. If $(d,1)/(d,0) \not\subseteq L_0$, then L_0 is distributive and hence $L \cong \mathbf{2}$. If $(d,1)/(d,0) \subseteq L_0$ then $L_0 = D_0[d]$ for a sublattice D_0 of D. But if $(d,1)/(d,0)$ is collapsed by f, then again $L \cong \mathbf{2}$. Suppose $(d,1)/(d,0)$ is not collapsed by f. Since L is subdirectly irreducible, and $f(d,1)/f(d,0)$ is critical, L is a smallest homomorphic image of $D_0[d]$ separating $(d,0)$ and $(d,1)$. By Lemma 4.29, the same holds for $D_0^*[d]$. Hence $L \cong D_0^*[d]$. Also D_0^* is a sublattice of D, and $D_0^*[d]$ is subdirectly irreducible, therefore Lemma 4.28 implies that $L \cong D_0^*[d]$ is isomorphic to $D'[d]$, where D' is a sublattice of D generated by d and a subset $X_d(D)$. □

Notice that there are at least $|X_d(D)| + 1$ nonisomorphic subdirectly irreducible members in $\{D[d]\}^V$, since if U, V are two subset of different cardinality, then $U \cup \{d\}$ and $V \cup \{d\}$ generate two nonisomorphic sublattices.

We now consider an interesting sequence of almost distributive lattices which is given in Lee [85], and was originally suggested by Jónsson.

Let K_i be the finite subdirectly irreducible almost distributive lattice represented by the $(i+1) \times (i+1)$ matrix $A(K_i)$ in Figure 4.17, and let $V_0 = \{K_1, K_2, K_3, \ldots\}^V$,

$$V_i = \{K_1, \ldots, K_{i-1}, K_{i+1}, \ldots\}^V \qquad \text{for } i = 1, 2, 3, \ldots.$$

LEMMA 4.31 $K_i \notin V_i$ for $i \in \{1, 2, 3 \ldots\}$.

PROOF. By Corollary 4.14 (vi) $K_i = D_i[d_i]$ for some finite distributive lattice D_i, $d_i \in D_i$, and K_i is a splitting lattice, so it generates a completely join prime variety for each i

(Lemma 2.8). Since $\mathcal{V}_i = \sum_{j\neq i}\{K_j\}^\mathcal{V}$ it suffices to show that $K_i \notin \{K_j\}^\mathcal{V}$ for any $i \neq j$. By the preceding lemma any subdirectly irreducible lattice in $\{K_j\}$ is isomorphic to a sublattice of $K_j = D_j[d_j]$ generated by a subset of $X_{d_j}(D_j)$. If $j < i$ then $|X_{d_j}(D_j)| < |X_{d_i}(D_i)|$ which certainly implies $K_i \notin \{K_j\}^\mathcal{V}$. Now suppose $j > i$ and let $X_{d_i}(D_i) = \{x_1, \ldots, x_{2i+2}\}$ with corresponding natural partitions

$$\{B_1, B_2, \ldots, B_{i+1}\} = \{\{x_1, x_2\}, \{x_3, x_4\}, \ldots, \{x_{2i+1}, x_{2i+2}\}\}$$
$$\{C_1, C_2, \ldots, C_{i+1}\} = \{\{x_2, x_3\}, \{x_4, x_5\}, \ldots, \{x_{2i+2}, x_1\}\}$$

and $X_{d_j}(D_j) = \{y_1, \ldots, y_{2j+2}\}$ with natural partitions

$$\{B_1', B_2', \ldots, B_{j+1}'\} = \{\{y_1, y_2\}, \{y_3, y_4\}, \ldots, \{y_{2j+1}, y_{2j+2}\}\}$$
$$\{C_1', C_2', \ldots, C_{j+1}'\} = \{\{y_2, y_3\}, \{y_4, y_5\}, \ldots, \{y_{2j+2}, y_1\}\}.$$

If f is an embedding of K_i into K_j, then we can assume without loss of generality that $f(x_1) = y_1$. As an embedding f must map B-blocks onto B'-blocks and C-blocks onto C'-blocks, hence $f(x_2) = y_2, \ldots, f(x_{2i+2}) = y_{2i+2}$. But $f(C_{i+1}) = \{f(x_{2i+2}), f(x_1)\} = \{y_{2i+2}, y_1\} \notin \{C_1', \ldots, C_{j+1}'\}$ which is a contradiction. Therefore K_i is not isomorphic to a sublattice of K_j, and consequently $K_i \notin \{K_j\}^\mathcal{V}$. $\square$

THEOREM 4.32 (Lee [85]). *Let $\mathcal{A}$ be the variety of all almost distributive lattices and let $\mathcal{V}_0$, $\mathcal{V}_i$ be defined as above.*

(i) *$|\Lambda_\mathcal{A}| = 2^\omega$.*

(ii) *There is an infinite descending chain of almost distributive varieties.*

(iii) *$\mathcal{V}_0$ has infinitely many dual covers.*

(iv) *There is an almost distributive variety with infinitely many covers in $\Lambda_\mathcal{A}$.*

PROOF. (i) By the preceding lemma, distinct subsets of $\{K_1, K_2, K_3, \ldots\}$ generate distinct subvarieties of $\mathcal{V}_0$.

(ii) Let $\mathcal{V}_i' = \{K_i, K_{i+1}, K_{i+2}, \ldots\}^\mathcal{V}$ for each $i \in \omega$. Then $\mathcal{V}_0 = \mathcal{V}_1' > \mathcal{V}_2' > \mathcal{V}_3' > \ldots$ follows again by Lemma 4.31.

(iii) We claim that K_i is the only finitely generated (hence finite) subdirectly irreducible member of $\mathcal{V}_0$ that is not in $\mathcal{V}_i$, from which it then follows that $\mathcal{V}_0 \succ \mathcal{V}_i$ for each $i \in \omega$.

By Lemma 4.31 $K_i \notin \mathcal{V}_i$. Every finite subdirectly irreducible member $L \in \mathcal{V}_0$ is a splitting lattice, so $L \in \{K_j\}^\mathcal{V}$ for some j. If $i \neq j$ then $L \in \mathcal{V}_i$, and if $L \in \{K_i\}^\mathcal{V}$ and L is not isomorphic to K_i then, by looking at the matrix that represents L, we see that $L \in \{K_j\}^\mathcal{V}$ for any $j > i$, so we also have $L \in \mathcal{V}_0$. This proves the claim.

(iv) Let $\overline{\mathcal{V}}_i$ be the conjugate variety of K_i relative to $\Lambda_\mathcal{A}$ ($i \in \omega$), and let $\overline{\mathcal{V}} = \bigcap_{i\in\omega} \overline{\mathcal{V}}_i$. We show that $\overline{\mathcal{V}} \prec \overline{\mathcal{V}} + \{K_i\}^\mathcal{V}$ for each i. By Theorem 2.3 (i) every subdirectly irreducible member of $\overline{\mathcal{V}} + \{K_i\}^\mathcal{V}$ belongs to $\overline{\mathcal{V}}$ or $\{K_i\}^\mathcal{V}$. Let L be a subdirectly irreducible lattice in $\{K_i\}^\mathcal{V}$. Lemma 4.30 implies that L is a sublattice of K_i, so $K_j \notin \{L\}^\mathcal{V}$ for any $j \neq i$. It follows that $L \in \overline{\mathcal{V}}$ or $L \cong K_i$, hence K_i is the only subdirectly irreducible lattice in $\overline{\mathcal{V}} + \{K_i\}^\mathcal{V}$ which is not in $\overline{\mathcal{V}}$. $\square$

4.4   Further Sequences of Varieties

In Section 4.3 we saw that above each of the varieties $\mathcal{L}_{13}$, $\mathcal{L}_{14}$ and $\mathcal{L}_{15}$ there is exactly one covering sequence of join irreducible varieties (Theorem 4.26). These results are due to Rose [84], and he also proved the corresponding results for $\mathcal{L}_6, \ldots, \mathcal{L}_{10}$. Since these varieties are not almost distributive, the proofs are more involved. Here we only consider the sequence $\mathcal{L}_6^n$ above $\mathcal{L}_6$.

Some technical results. Let L be a lattice and X a subset of L. An element $z \in L$ is said to be *X-join isolated* if $z = x + y$ and $x, y < z$ implies $x, y \in X$. The notion of an *X-meet isolated* element is defined dually. A quotient u/v of L is said to be *isolated* if every element of u/v is u/v-join isolated and u/v-meet isolated.

The next four lemmas (4.32–4.35) appear in Rose [84], where they are used to prove that the variety $\mathcal{L}_i^{n+1}$ is the only join irreducible cover of $\mathcal{L}_i^n$ for $i \in \{6,7,8,9,10\}$ (see Figure 2.2). These lemmas only apply to lattices satisfying certain conditions summarized here as

CONDITION $(*)$. L is a finite subdirectly irreducible neardistributive lattice with critical quotient c/a (which is unique by Theorem 4.8). Furthermore c'/a' is a quotient of L such that

(i) $a' \le a < c \le c'$;

(ii) any $z \in c'/a' - \{a'\}$ is c'/a'-join isolated;

(iii) any $z \in c'/a' - \{c'\}$ is c'/a'-meet isolated.

Observe that if $b \notin c'/a'$ and b is noncomparable with some $z \in c'/a'$, then b is noncomparable with all the elements of c'/a'. Moreover, $a' + b = z + b = c' + b$ and $a'b = zb = c'b$, which implies $N(c'/a', b)$. Hence, for any $b \notin c'/a'$, the conditions $N(c'/a', b)$ and $N(c/a, b)$ are equivalent.

LEMMA 4.33 *Suppose L satisfies condition $(*)$.*

(i) *If $u \succ c'$ in L, then there exists $b \in L$ such that $N(c'/a', b)$ and $u = a' + b \succ b \succ a'b$.*

(ii) *If L excludes L_7, then we also have $a' \succ a'b$.*

PROOF. (i) By Lemma 4.16 (iii) there exists $b \in L$ such that $N(c/a, b)$, $b \le u$, $b \not\le c'$ and $u/c' \searrow_\beta a + b/(a + b)c'$. b is noncomparable with c', so $N(c', a', b)$ follows from the remark above, and we cannot have $(a + b)c' < c'$, since $(a + b)c'$ is not c'/a'-meet isolated. So $(a + b)c' = c'$ and therefore $u = a + b = a' + b$. Since L is finite we can choose t such that $b \le t \prec a' + b$. t is also noncomparable with c', so we get $N(c'/a', t)$, and of course $u = a' + t$. Hence we may assume that $u \succ b$. Also $b \succ a'b$, since $a'b < t < b$ would imply $N(b/t, a')$, hence $N(b/t, a)$, and by Corollary 4.7 $N(c/a, a)$, which is impossible.

(ii) Suppose to the contrary, that $a'b < t \prec a'$ for some $t \in L$. By the dual of part (i) there exists $b_0 \in L$ with $N(c'/a', b_0)$ and $t = a'b_0 \prec b_0 \prec a' + b_0$ (Figure 4.18 (i)). Since $a' + b \succ b$, we have $t + b = a' + b$ and so $N(a'/t, b)$. Now $a'/t \nearrow a' + b_0/b_0$ and Corollary 4.7 imply $N(a' + b_0/b_0, b)$. Thus $a' + b_0 \not\ge a' + b$, which clearly implies that b_0 and $a' + b_0$ are noncomparable with $a' + b$. Since $a' + b \succ c$, $b \succ a'b$ and $b_0 \succ y$, we must have $(a' + b)(a' + b_0) = c'$, $(a' + b_0)b = a'b$ and $(a' + b)b_0 = t$. Hence the elements a', b and b_0 generate L_7 (Figure 4.18 (i)), and this contradiction completes the proof. □

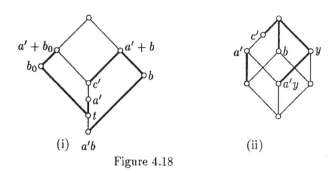

Figure 4.18

We now add the following condition.

CONDITION (**). b is an element of L such that $N(c'/a', b)$ and $a' + b/a'b = \{b, a' + b, a'b\} \cup c'/a'$.

LEMMA 4.34 *If L satisfies conditions* (*), (**) *and excludes L_{14}, then for $x, y \in L$,*

(i) $a' + b = a' + y > y$ *implies* $y \leq b$;

(ii) $a' + b = x + b > x$ *implies* $x \leq c'$.

PROOF. (i) If $y \nleq b$, then y is noncomparable with b and with a'. We claim that y can be chosen so that a', c', b, y generate L_{14} (see Figure 4.18 (ii)). We may assume that $y \prec a' + y$. If $a'y < t < y$, then we would have $N(y/t, a')$, hence $N(y/t, a)$, and by Corollary 4.7 $N(c/a, a)$, which is impossible. Therefore $y \succ a'y$. By semidistributivity

$$a' + b = a' + y = b + y = a'b + a'y + by.$$

From this it follows that the elements $a'b = c'b$, $a'y = c'y$ and by are noncomparable, and therefore

$$a' = a'b + a'y, \qquad b = a'b + by \quad \text{and} \quad y = a'y + by.$$

This shows that a', b and y generate an eight element Boolean algebra. Since $N(c'/a', b)$ and $N(c'/a', y)$ hold, L includes L_{14}.

(ii) If $x \nleq c$, then $a' + x = a' + b$, and since we cannot have $x \leq b$, part (i) implies that L includes L_{14}. □

LEMMA 4.35 *If L satisfies conditions* (*), (**) *and excludes L_7, L_{13} and L_{15}, then c' is meet irreducible.*

PROOF. Suppose c' is meet reducible. Then there exists an element x covering c' such that $c' = x(a' + b)$. By Lemma 4.33 (i) there exists $b_0 \in L$ with $N(c'/a', b_0)$ and $x = a' + b_0 \succ b_0 \succ a'b_0$. The elements $a' + b$, $a' + b_0$ and $b + b_0$ generate a lattice K that is a homomorphic image of the lattice in Figure 4.19 (i). If K is isomorphic to that lattice, then $bb_0 \leq a'$, since $bb_0 \nleq a'$ would imply $bb_0 + a' \in c'/a' - \{a'\}$, contradicting the

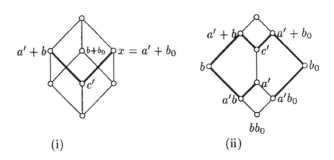

Figure 4.19

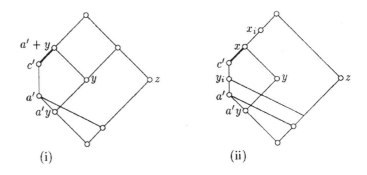

Figure 4.20

assumption that every element of $c'/a' - \{a'\}$ is c'/a'-join isolated. In fact we must have $bb_0 < a'$, since a' is c'/a'-meet isolated. Thus $K \cup \{a'\}$ is a sublattice of L isomorphic to L_{13}, contrary to the hypothesis. We infer that K is a proper homomorphic image of the lattice in Figure 4.19 (i), and since $a' + b$, $a' + b_0$ and c' are distinct, it follows that $c' < b + b_0$. Now Figure 4.19 (ii) shows that if $a'b$ and $a'b_0$ are noncomparable, then L includes L_{15}, while $a'b < a'b_0$ or $a'b_0 < a'b$ imply that L includes L_7. Finally, we cannot have $a'b = a'b_0$, since then L includes L_1, which contradicts the semidistributivity of L.□

LEMMA 4.36 *If L satisfies conditions* $(*), (**)$ *and excludes L_9, L_{13}, L_{14} and L_{15} then b is meet irreducible.*

PROOF. To avoid repetition, we first establish two technical results:

(A) If $N(u/v, z)$ for some $u/v \not\subseteq c'/a'$ and $z \notin c'/a'$, then there exists $y \in L$ with $N(c'/a', y)$ such that $N(a' + y/a', z)$, $N(y/a'y, z)$ and $a' + y \succ c'$ (Figure 4.20 (i)) or dually.

Consider a sequence $u/v = x_0/y_0 \sim_w x_1/y_1 \sim_w \ldots \sim_w x_n/y_n = c/a$. Since c/a is a subquotient of c'/a' and u/v is not, there is an index $i > 0$ such that $x_i/y_i \not\subseteq c'/a'$ and

$x_{i+1}/y_{i+1} \subseteq c'/a'$. By duality, suppose that $x_i/y_i \searrow_w x_{i+1}/y_{i+1}$. Since y_{i+1} is c'/a'-meet isolated, $y_i \in c'/a'$, and since $x_i c' < c'$ would also imply $x_i \in c'/a'$, we must have $x_i c' = c'$, and therefore $x_i > c' \geq y_i$. Now Lemma 4.6, and the fact that u/v projects weakly onto x_i/y_i and c/a imply $N(x_i/y_i, z)$ and $N(c/a, z)$. Since $z \not\in c'/a'$ we must have $N(c'/a', z)$. Choose $x \in L$ such that $c' \prec x \leq x_i$, then clearly $N(x/a', z)$ holds (Figure 4.20 (ii)). By Lemma 4.33 (i) there exists $y \in L$ with $N(c'/a', y)$ and $x = a' + y$. Since $x/a' \searrow y/a'y$, Lemma 4.6 again implies $N(y/a'y, z)$. This proves (A).

(B) *If for some* $u, v, z \in L$ *with* $z \geq b$ *we have* $N(u/v, z)$, *then* $u/v \subseteq c'/a'$.

Suppose $u/v \not\subseteq c'/a'$. Since clearly $z \in c'/a'$, (A) implies that there exists $b_0 \in L$ such that $n(c'/a', b_0)$ and either

(1) $N(a' + b_0/a', z)$, $N(b_0/a'b_0, z)$ and $c' \prec a' + b_0$ or

(2) $N(c'/a'b_0, z)$, $N(a' + b_0, z)$ and $a'b_0 \prec a'$.

We will show that, contrary to the hypothesis of the lemma, the elements a', c', b and b_0 generate L_{15}. Since we already know that $N(c'/a', b)$ and $N(c'/a', b_0)$, it suffices to check that $a'b + a'b_0 = a'$ and $(a' + b)(a' + b_0) = c'$. Either of (1) or (2) imply that z is noncomparable with $a'b_0$ and $a' + b_0$. Since $a'b < b \leq z$ we must have $a'b_0 \not\leq a'b$. Strict inclusion $a'b < a'b_0$ is also not possible, because $a'b \prec a'$ and $a'b_0 < a'$. Thus $a'b$ and $a'b_0$ are noncomparable, and since $a'b \prec a'$, it follows that $a'b + a'b_0 = a'$. Next note that $a'z = a'b$, because $a'b \prec a'$ and $a'b \leq a'z < a'$. Hence $a'z = a'b > a'bb_0 = a'b_0 z$, so we cannot have $N(c'/a'b_0, z)$ in (2). Therefore (1) must hold, and in particular $N(a'+b_0/a', z)$, whence it follows that $a' + b_0 \not\geq b$. Thus $c' \leq (a'+b)(a'+b_0) < a' + b$, and since $c' \prec a'+b$, $c = (a' + b)(a' + b_0)$. This proves (B).

Proceeding now with the proof of the lemma, suppose b is meet reducible. Then we can find $z \succ b$ such that $b = (a' + b)z$. Consider a shortest sequence

$$z/b = x_0/y_0 \sim_w x_1/y_1 \sim_w \ldots \sim_w x_n/y_n = c/a.$$

Clearly $n \geq 2$. The case $n = 2$ can also be ruled out, since z/b is a transpose of x_1/y_1, while Theorem 4.8 implies that c/a is a subquotient of x_{n-1}/y_{n-1}, hence $x_1/y_1 = x_{n-1}/y_{n-1}$ would imply $z = x_1 + b \geq a' + b$ or $b = y_1 z \leq ab$, both of which are impossible. Thus $n \geq 3$. If $z/b \nearrow x_1/y_1 \searrow x_2/y_2$, then x_1/y_1 is prime, since $y_1 < t < x_1$ would imply $N(t/y_1, z)$, and by (B) $t/y_1 \subseteq c'/a'$, which leads to a contradiction, as $b \not\leq c'$. Similarly x_2/y_2 must be prime, because $y_2 < t < x_2$ would imply $N(x_2, t, y_1)$, whence (B) gives $x_2/t \subseteq c'/a'$. This contradicts the semidistributivity of L, since $x_1 = y_1 + z = y_1 + x_2$, but $x_2 z = c'z < y_1$. Hence $z/b \nearrow_\beta x_1/y_1 \searrow_\beta x_2/y_2$, and now Lemma 4.4 implies that the sequence can be shortened, contrary to our assumption. Consequently we must have $z/b \searrow x_1/y_1 \nearrow x_2/y_2$. Observe that $x_1 \not\leq c'/a'$, for otherwise $z = x_1 + b = a' + b$. Again the quotient x_1/y_1 is prime, since $y_1 < t < x_1$ would imply $N(x_1/t, b)$, contradicting (B). However x_2/y_2 cannot be prime because of the minimality of n. So there exists $u \in L$ with $y_2 < v < u < x_2$ such that $N(u/v, x_1)$ holds (Figure 4.21 (i)). By Corollary 4.7 we have $N(c/a, x_1)$, and since $x_1 \not\leq c'/a'$, $N(c'/a', x_1)$ holds. Notice that $y_1 = (a' + b)zx_1 = (a' + b)x_1 \geq a'x_1$. We claim that $y_1 = a'x_1$. Suppose to the contrary that $a'x_1 < y_1$. Then $u/v \not\subseteq c'/a'$ since $ux_1 = y_1 \neq a'x_1$. By (A) there exists $b_0 \in L$ with $N(c'/a', b_0)$ such that $N(a' + b_0/a', x_1)$ and $a' + b_0 \succ c'$, or dually $N(c'/a'b_0, x_1)$ and $a'b_0 \prec a'$.

First suppose that $N(a' + b_0/a', x_1)$. We cannot have $a' + b_0 < a' + b$ since $a' + b \succ c$. On the other hand $a' + b \leq a' + b_0$ implies $N(a' + b/a', x_1)$, whence $a'x_1 = (a' + b)x_1 = y_1$,

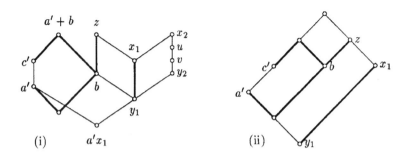

Figure 4.21

a contradiction. Therefore $a' + b$ and $a' + b_0$ are noncomparable and $(a' + b)(a' + b_0) = c'$. Since L excludes L_{13} and L_{15}, it follows as in the proof of Lemma 4.35 that the a', c', b and b_0 generate L_7. Thus $a' + b < b + b_0$, and as $a' \succ a'b$, we can only have $a'b_0 < a'b$. By Lemma 4.6 $N(a' + b_0/a', x_1)$ and $a' + b_0/a' \searrow b_0/a'b_0$ imply $N(b_0/a'b, x_1)$. Hence $a'b_0 + x_1 = b_0 + x_1$, and together with $a'b_0$ and $x_1 \leq z$ this implies $b_0 \leq b_0 + x_1 = a'b_0 + x_1 \leq z$. It follows that $a' + b < b + b_0 \leq z$, which is a contradiction.

Now suppose that $N(c'/a'b_0, x_1)$. Since we are also assuming that L excludes L_{14}, we can dualize the above argument to again obtain a contradiction. Thus $y_1 = a'x_1$. We complete the proof by showing that a', c', b and x_1 generate L_9 (Figure 4.21 (ii)). Clearly $a' \geq a'x_1 = y_1$ implies $a'b \geq y_1b = y_1$. In fact we must have $a'b > y_1$, since $a'b = y_1 = a'x_1 < x_1$ would imply $x_1 \geq b$ by the dual of Lemma 4.34 (i), a contradiction. Also $a'(a'b + x_1) = a'b < a'b + x_1$, since $a' \succ a'b$, and now the dual of Lemma 4.34 (i) implies $a'b + x_1 \geq b$. Hence $a'b + x_1 = b + x_1 = z$. Finally $a' + b/c' \searrow b/a'b$, $N(b/a'b, x_1)$ and Lemma 4.6 imply $N(a' + b/c', x_1)$, whence $a'x_1 = (a' + b)x_1$.   □

The sequence $\mathcal{L}_6^n$. The next theorem is in preparation to proving the result due to Rose [84] that $\mathcal{L}_6^{n+1}$ is the only join irreducible cover of $\mathcal{L}_6^n$. A quotient c/a of a lattice is an L_6^n-quotient if for some $b, b_0, \ldots, b_n \in L$ the set $\{a, c, b, b_0, \ldots, b_n\}$ generates a sublattice of L isomorphic to L_6^n, with c/a as critical quotient (Figure 2.2). In this case we shall write $L_6^n(c/a, b, b_0, \ldots, b_n)$.

THEOREM 4.37 (Rose [84]). *Let L be a subdirectly irreducible lattice, and assume that the variety $\{L\}^\mathcal{V}$ contains none of the lattices $M_3, L_1, \ldots, L_5, L_7, \ldots, L_{15}$. Suppose further that, for some $k \in \omega$, c/a is an L_6^k-quotient of L. Then*

   (i) *if L does not have any L_6^{k+1}-quotients, then c/a is a critical quotient of L and $L/\mathrm{con}(a, c)$ has no L_6^k-quotients.*

   (ii) *if L is finite and $L \not\cong L_6^k$, then c/a is an L_6^{k+1}-quotient.*

PROOF.(i) By Theorem 4.1 $\{L\}^\mathcal{V}$ is semidistributive, and by Theorem 4.8 L has a unique critical quotient, which we denote by x/y. Choose $b, b_0, \ldots, b_k$ so that $L_6^k(c/a, b, b_0, \ldots, b_k)$

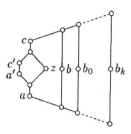

Figure 4.22

holds. We will prove several statements, the last of which shows that $x/y = c/a$. The first three are self-evident.

(A) *Any nontrivial subquotient c'/a' of c/a is an L_6^k-quotient.*

(B) *Suppose that for some $a', c', z \in L$ we have $N(c'/a', z)$ with $a \leq a'z < a' + z \leq c$. Then $L_6^{k+1}(c'/a', z, b, b_0, \ldots, b_k)$ holds (see Figure 4.22).*

(C) *Suppose that for some $z \in L$ we have $N(a + b_i/ab_i, z)$ $(i \in \{0, \ldots, k\})$. Then $L_6^{i+1}(c/a, b, b_0, \ldots, b_i, z)$ holds, and similarly if $N(a + b/ab, z)$ then we have $L_6^0(c/a, b, z)$.*

(D) *For any quotients u/v and p/q in L, if $u/v \nearrow p/q \searrow c/a$, then $u/v \searrow uc/va \nearrow c/a$, and all four transpositions are bijective.*

By Lemma 4.5 the lattice generated by q, c, b is a homomorphic image of the lattice in Figure 4.23 (i). The pentagon $N(r/d, b)$ is contained in $a + b/ab$, whence it follows that $L_6^k(r/d, b, b_0, \ldots, b_k)$. From this we infer that r/d is distributive, for otherwise r/d would contain a pentagon $N(c'/a', b')$ (by semidistributivity L excludes M_3), and we would have

$$L_6^{k+1}(c'/a', b', b, b_0, \ldots, b_k).$$

Hence the transposition $r/s \searrow e/d$ is bijective. By Lemma 4.6, the transpositions $p/q \searrow r/s$ and $e/d \searrow c/a$ are also bijective, and we consequently have $p/q \searrow_\beta c/a$.

Again by Lemma 4.5, the lattice generated by q, u, b is a homomorphic image of the lattice in Figure 4.23 (ii). Note that $ab \leq bq < b \leq a + b$, whence $N(b/bq, b_0)$. Since $v + b/d' \searrow b/bq$, it follows by still another application of Lemma 4.5 that the lattice generated by d', b and b_0 is a homomorphic image of the lattice in Figure 4.23 (iii) and by Lemma 4.12 the transposition $v + b/d' \searrow r''/s''$ is bijective. Put $t = r'(b + b_0)$ to obtain $N(t/s'', b)$ and therefore $L_6^k(t/s, b, b_0, \ldots, b_k)$. This implies that t/s'' is distributive, and so is r'/d', since the two quotients are isomorphic. The transposition $e'/d' \nearrow r'/s'$ is therefore bijective, and the bijectivity of $u/v \nearrow e'/d'$ and $r'/s' \nearrow p/q$ follows from Lemma 4.12. Consequently $u/v \nearrow_\beta p/q$. Now semidistributivity (Lemma 4.4) implies $u/v \searrow uc/va \nearrow c/a$. By duality, these two transpositions must also be bijective.

(E) *If c/a projects weakly onto a quotient u/v, then $u/v \searrow uc'/va' \nearrow c'/a'$ for some subquotient c'/a' of c/a*

Assume that $c/a = x_0/y_0 \sim_w x_1/y_1 \sim_w \ldots \sim_w x_n/y_n = u/v$, where the transpositions alternate up and down. We use induction on n. The cases $n = 0, 1$ are trivial, so by duality we may assume that $c/cy_1 \nearrow x_1/y_1 \supseteq y_1 + x_2/y_1 \searrow x_2/y_2$. Since c/cy_1 is also

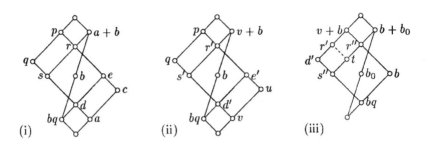

Figure 4.23

a L_6^k-quotient we can apply (D) to conclude that the first transpose must be bijective. Hence $y_1 + x_2/y_1$ transposes bijectively onto a subquotient c'/a' of c/a ($a' = cy_1$). A second application of (D) gives $c'/a' \searrow c'x_2/a'y_2 \nearrow x_2/y_2$, proving the case $n = 2$, while for $n > 2$ the sequence can now be shortened by one step. The result follows by induction.

(F) $x/y = c/a$.

Since x/y is critical and prime, c/a projects weakly onto x/y. By (E) x/y projects onto a subquotient c'/a' of c/a and since x/y is the only critical quotient of L, we must have $x/y = c'/a'$. If $x < c$, then the hypothesis of part (i) (of the theorem) is satisfied with a replaced by x, and we infer that x/y is a subinterval of c/x, which is impossible. Hence $x = c$, and similarly $y = a$, which also shows that c/a is the only L_6^k-quotient of L.

To complete the proof of part (i), suppose $\overline{L} = L/\mathrm{con}(a,c)$ contains an L_6^k-quotient, i.e. for some $u, v, d, d_0, \ldots, d_k \in L$ we have $L_6^k(\overline{u}/\overline{v}, \overline{d}, \overline{d}_0, \ldots, \overline{d}_k)$ in $\overline{L}$. If $c = u$ in L, then $u = c > a > v$ and $L_6^k(u/v, d, d_0, \ldots, d_k)$, which would contradict the fact that c/a is the only L_6^k-quotient of L. Thus $c \neq u$ and, similarly, $a \neq u$ and $c, a \neq v$. If $a = d$, then we must have $N(u/v, a)$ in L. But by Corollary 4.7 this would imply $N(c/a, a)$, which is impossible. So $a \neq d$ and, more generally, $c, a \notin \{d, d_0, \ldots, d_k\}$. Since $\mathrm{con}(a,c)$ identifies only a and c, we infer that $L_6^k(u/v, d, d_0, \ldots, d_k)$ in L with $u/v \neq c/a$, and this contradiction concludes part (i).

For the proof of part (ii), we will use the concept of an isolated quotient and all its implications (Lemmas 4.33 – 4.36). Let c'/a' be an isolated quotient of L such that $c/a \subseteq c'/a'$.

(G) *Suppose that for some* $b \in L$ *we have* $N(c'/a', b)$ *with* $a'b \prec a'$, $c' \prec a' + b$ *and* $a'b \prec b \prec a' + b$. *Then*

(1) $a' + b/a'b = c'/a' \cup \{a'b, b, a' + b\}$;

(2) $a' + b/a'b$ *is an isolated quotient of* L.

Assume (1) fails. Then there exists $x \in L$ such that $x \in a' + b/a'b$ but $x \notin c'/a' \cup \{a'b, b, a' + b\}$. Since $a'b < x < a' + b$ and $a'b \prec b \prec a' + b$, it follows that b and x are noncomparable and $xb = a'b$, $x + b = a' + b$. Furthermore, as c'/a' is isolated, x is noncomparable with a' and c', whence $a' + x = a' + b$. This, however, contradicts the semidistributivity of L, since $a' + b \neq a' + xb = a'$. Therefore (1) holds.

To prove (2), it suffices to show that

(3) a' is join irreducible and c' is meet irreducible;

(4) b is both join and meet irreducible;

(5) $x \in L$ and $a' + b = x + b > x$ imply $x \in c'/a'$;

(6) $y \in L$ and $a' + b = a' + y > y$ imply $y = b$;

(7) $x \in L$ and $a'b = xb < x$ imply $x \in c'/a'$;

(8) $y \in L$ and $a'b = a'y < y$ imply $y = b$.

(3) and (4) follow from Lemmas 4.35 and 4.36 and their duals respectively. Suppose $a' + b = x + b > x$. Then $x \le c'$ by Lemma 4.34 (ii) and, since $x + b \ne b$, we have $x \not\le c'b = a'b$. Now $x \not\prec a'$, because $a'b$ is the only dual cover of a'. Since c'/a' is isolated, this implies $x \in c'/a'$, whence (5) holds. If $a' + b = a' + y$, then Lemma 4.34 (i) implies that $y \le b$, and from the join irreducibility of b we infer $y = b$, thereby proving (6). Finally, (7) and (8) are the duals of (5) and (6).

(H) If $L_6^k(c/a, b, b_0, \ldots, b_k)$, then the elements $b, b_0, \ldots, b_k \in L$ can be chosen such that

$$a + b/ab = c/a \cup \{ab, b, a + b\},$$
$$a + b_0/ab_0 = a + b/ab \cup \{ab_0, b_0, a + b_0\},$$
$$a + b_i/ab_i = a + b_{i-1}/ab_{i-1} \cup \{ab_i, b_i, a + b_i\} \qquad \text{for} \quad i \in \{1, 2, \ldots, k\},$$

and all these quotients are isolated.

By Lemma 4.35 and its dual, the quotient c/a is isolated. Choose $x \in L$ with $c \prec x \le a + b$. Since L excludes L_7, Lemma 4.33 (i), (ii) and (G) above imply the existence of $b' \in L$ with $N(c/a, b')$ and $a + b' = x$, such that this sublattice is an interval in L and is isolated. Since a is join irreducible and $a \prec ab'$, we infer that $ab' \ge ab$. Thus $ab \le ab' \prec b' \prec x \le a + b$, whence it follows that $L_6^k(c/a, b', b_0, \ldots, b_k)$. So we may replace b by b', and continuing in this way we prove (H).

Since c/a is a prime L_6^k-quotient of L, (H) implies that we can find $b, b_0, \ldots, b_k$ in L such that the sublattice generated by c/a and these b's is an interval of L. Since $L \cong L_6^k$, there exists $u \in L$ with $u \succ a + b_k$ or $u \prec ab = k$ and from Lemma 4.33 (i) or its dual we obtain $b' \in L$ such that $N(a + b_k/ab_k, b')$, which implies $L_6^{k+1}(c/a, b, b_0, \ldots, b_k, z)$ as required. □

After much technical detail we can finally prove:

THEOREM 4.38 (Rose [84]). *$\mathcal{L}_6^{n+1}$ is the only join irreducible cover of $\mathcal{L}_6^n$.*

PROOF. Suppose to the contrary that for some natural number n, the variety $\mathcal{L}_6^n = \{L_6^n\}^{\mathcal{V}}$ has a join irreducible cover $\mathcal{V} \ne \mathcal{L}_6^{n+1}$. Choose n as small as possible. Since $\mathcal{V}$ has finite height in Λ, it is completely join irreducible, so it follows from Theorem 2.5 that $\mathcal{V} = \{L\}^{\mathcal{V}}$ for some finitely generated subdirectly irreducible lattice L. Note that $L_6^n \in \{L\}^{\mathcal{V}}$

Using the results of Section 2.3 one can check that L_6^n is a splitting lattice, and since it also satisfies Whitman's condition (W), Theorem 2.19 implies that L_6^n is projective in $\mathcal{L}$. By Lemma 2.10 L_6^n is a sublattice of L, so for some $a, c, b, b_0, \ldots, b_n \in L$ we have $L_6^n(c/a, b, b_0, \ldots, b_n)$. By Theorem 4.37 (i) c/a is critical, and $L/\text{con}(a, c)$ has no L_6^n-quotients. Again, since L_6^n is subdirectly irreducible and projective, Lemma 2.10 implies that L_6^n is not a member of the variety generated by $L/\mathrm{con}(a,c)$. This, together with the minimality of n implies that, for $n = 0$, $L/\mathrm{con}(a,c)$ is a member of $\mathcal{N}$ and, for $n > 0$, $L/\mathrm{con}(a,c)$ is in $\mathcal{L}_6^{n_1}$. By Lemma 4.15 L is finite and, since $L \ncong L_6^n$, it follows from Theorem 4.38 (ii) that L includes L_6^{n+1}. This contradiction completes the proof. □

By a similar approach Rose [84] proves that $\mathcal{L}_i^{n+1}$ is the only join irreducible cover of $\mathcal{L}_i^n$ for $i = 7$ and 9 (the cases $i = 8$ and 10 follow by duality). A slight complication arises due to the fact that L_7^n and L_9^n are not projective for $n \geq 1$, since the presence of doubly reducible elements implies that (W) fails in these lattices. As a result the final step requires an inductive argument. For the details we refer the reader to the original paper of Rose [84].

Further results about nonmodular varieties. The variety $\mathcal{M}_3 + \mathcal{N}$ is the only join reducible cover of $\mathcal{N}$ (and $\mathcal{M}_3$), and its covers have been investigated by Ruckelshausen [78]. His results show that the varieties $\mathcal{V}_1, \ldots, \mathcal{V}_8$ generated by the lattices $V_1, \ldots, V_8$ in Figure 2.4 are the only join irreducible covers of $\mathcal{M}_3 + \mathcal{N}$ that are generated by a planar lattice of finite length.

The techniques used in the preceding investigations make extensive use of Theorem 4.8, and are therefore unsuitable for the study of varieties above $\mathcal{L}_{11}$ or $\mathcal{L}_{12}$. Rose [84] showed that $\mathcal{L}_{12}$ has at least two join irreducible covers, generated by the two subdirectly irreducible lattices L_{12}^1 and G respectively, (see Figure 4.24, dual considerations apply to $\mathcal{L}_{11}$).

Using methods developed by Freese and Nation [83] for the study of covers in free lattices, Nation [85] proves that these are the only join irreducible covers of $\mathcal{L}_{12}$, and that above each of these is exactly one covering sequence of join irreducible varieties $\mathcal{L}_{12}^n$ and $\mathcal{G}^n = \{G^n\}$ (Figure 4.24).

By a result of Rose [84], any semidistributive lattice which fails to be bounded contains a sublattice isomorphic to L_{11} or L_{12} (see remark after Theorem 4.10). Thus it is interesting to note that the lattices L_{12}^n and G^n are again splitting lattices.

In Nation [86] similar techniques are used to find a complete list of covering varieties of $\mathcal{L}_1$ (and $\mathcal{L}_2$ by duality). The ten join irreducible covers are generated by the subdirectly irreducible lattices $L_{16}, \ldots, L_{25}$ in Figure 4.25.

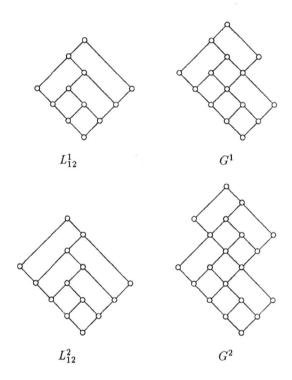

L_{12}^1
G^1

L_{12}^2
G^2

Figure 4.24

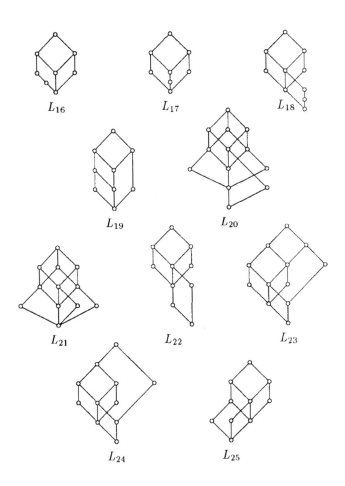

Figure 4.25

Chapter 5

Equational Bases

5.1 Introduction

An *equational basis* for a variety $\mathcal{V}$ of algebras is a collection $\mathcal{E}$ of identities such that $\mathcal{V} = \text{Mod}\mathcal{E}$. An interesting problem in the study of varieties is that of finding equational bases. Of course the set $\text{Id}\mathcal{V}$ of all identities satisfied by members of $\mathcal{V}$ is always a basis, but this set is generally highly redundant, so we are interested in finding proper (possibly minimal) equational basis for $\mathcal{V}$. In particular we would like to know under what conditions $\mathcal{V}$ has a *finite* equational basis.

It might seem reasonable to conjecture that every finitely generated variety is finitely based, but this is not the case in general. Lyndon [54] constructed a seven-element algebra with one binary operation which generates a nonfinitely based variety, and later a four-element and three-element example were found by Višin [63] and Murskiĭ [65] respectively. On the other hand Lyndon [51] proved that any two element algebra with finitely many operations does generate a finitely based variety. The same is true for finite groups (Oates and Powell [64]), finite lattices (even with finitely many additional operations, McKenzie [70]), finite rings (Kruse [77], Lvov [77]) and various other finite algebras.

Shortly after McKenzie's result, Baker discovered that any finitely generated congruence distributive variety is finitely based. Actually his result is somewhat more general and moreover, the proof is constructive, meaning that for a particular finitely generated congruence distributive variety one can follow the proof to obtain a finite basis. However the proof, which only appeared in its final version in Baker [77], is fairly complicated and several nonconstructive shortcuts have been published (see Herrmann [73], Makkai [73], Taylor [78] and also Burris and Sankappanavar [81]). The proof that is presented in this chapter is due to Jónsson [79] and is a further generalization of Baker's theorem.

In contrast to these results on finitely based lattice varieties, McKenzie [70] gives an example of a lattice variety that is not finitely based. Another example by Baker [69], constructed from lattices corresponding to projective planes, shows that there is a nonfinitely based modular variety.

Clearly an equational basis for the meet (intersection) of two varieties is given by the union of equational bases for the two varieties, which implies that the meet of two finitely based varieties is always finitely based. An interesting question is whether the same is true for the join of two finitely based varieties. This is not the case, as was independently discovered by Jónsson [74] and Baker. The example given in Baker [77'] is included in this chapter and actually shows that even with the requirement of modularity the above question has a negative answer. In Jónsson's paper, however, we find sufficient conditions for a positive answer and these ideas are generalized further by Lee [85']. One consequence is that the join of the variety $\mathcal{M}$ of all modular lattices and the smallest nonmodular variety $\mathcal{N}$ is finitely based. This variety, denoted by $\mathcal{M}^+(= \mathcal{M} + \mathcal{N})$, is a cover of $\mathcal{M}$, and an equational basis for $\mathcal{M}^+$ consisting of just eight identities is presented in Jónsson [77].

Recently Jónsson showed that the join of two finitely based modular varieties is finitely based whenever one of them is generated by a lattice of finite length. A generalization of this result and further extensions to the case where one of the varieties is nonmodular appear in Kang [87].

Although Baker's theorem allows one to construct, in principle, finite equational bases for any finitely generated lattice variety, the resulting basis is usually too large to be of any practical use. In Section 5.4 we give some examples of finitely based varieties for which reasonably small equational bases have been found. These include the varieties $\mathcal{M}_n$ ($n \in \omega$, from Jónsson [68]), $\mathcal{N}$ (McKenzie [72]) and the variety $\mathcal{M}^+$ referred to above.

5.2 Baker's Finite Basis Theorem

Some results from model theory. A class $\mathcal{K}$ of algebras is an *elementary class* if it is the class of all algebras which satisfy some set $\mathcal{S}$ of first-order sentences (i.e. $\mathcal{K} = \text{Mod}\,\mathcal{S}$), and $\mathcal{K}$ is said to be *strictly elementary* if $\mathcal{S}$ may be taken to be finite or, equivalently, if $\mathcal{K}$ is determined by a single first-order sentence (the conjunction of the finitely many sentences in $\mathcal{S}$).

(These concepts from model theory are applicable to any class of models of some given first-order language. Here we assume this to be the language of the algebras in $\mathcal{K}$. For a general treatment consult Chang and Keisler [73] or Burris and Sankappanavar [81].)

The problem of finding a finite equational basis is a particular case of the following more general question: When is an elementary class strictly elementary?

Recall the definition of an ultraproduct from Section 1.3. The nonconstructive short-cuts to Baker's finite basis theorem make use of the following well-known result about ultraproducts:

THEOREM 5.1 (Los[55]). *Let* $A = \prod_{i \in I} A_i$ *and suppose* $\phi_{\mathcal{U}}$ *is the congruence induced by some ultrafilter* $\mathcal{U}$ *over the index set* I. *Then, for any first-order sentence* σ, *the ultraproduct* $A/\phi_{\mathcal{U}}$ *satisfies* σ *if and only if the set* $\{i \in I : A_i$ *satisfies* $\sigma\}$ *is in* $\mathcal{U}$.

In particular this theorem shows that elementary classes are closed under ultraproducts. But it also has many other consequences. For example we can deduce the following two important results:

THEOREM 5.2 (Frayne, Morel and Scott [62], Kochen [61]). *An elementary class* $\mathcal{K}$ *of algebras is strictly elementary if and only if the complement of* $\mathcal{K}$ *is closed under ultraproducts. The complement can be taken relative to any strictly elementary class containing* $\mathcal{K}$.

PROOF. Suppose $\mathcal{B}$ is an elementary class that contains $\mathcal{K}$. If $\mathcal{K}$ is strictly elementary, then membership in $\mathcal{K}$ can be described by a first-order sentence. By the preceding theorem the negation of this sentence is preserved by ultraproducts, so any ultraproduct of members in $\mathcal{B} - \mathcal{K}$ must again be in $\mathcal{B} - \mathcal{K}$.

Conversely, suppose $\mathcal{K}$ is elementary and is contained in a strictly elementary class $\mathcal{B}$. Assuming that $\mathcal{B} - \mathcal{K}$ is closed under ultraproducts, let $\mathcal{S}$ be the set of all sentences that hold in every member of $\mathcal{K}$, and let I be the collection of all finite subsets of $\mathcal{S}$. Since $\mathcal{B}$ is strictly elementary, $\mathcal{B} = \text{Mod}\,\mathcal{S}_0$ for some $\mathcal{S}_0 \in I$.

If $\mathcal{K}$ is not strictly elementary then, for each $i \in I$, there must exist an algebra A_i not in $\mathcal{K}$ such that A_i satisfies every sentence in the finite set $i \cup \mathcal{S}_0$. Note that this implies $A_i \in \mathcal{B} - \mathcal{K}$. We construct an ultraproduct $A/\phi_{\mathcal{U}} \in \mathcal{K}$ as follows:

Let $A = \prod_{i \in I} A_i$ and, for each $i \in I$ define $J_i = \{j \in I : j \supseteq i\}$. Then $J_i \neq \emptyset$ and $J_i \cap J_k = J_{i \cup k}$ for all $i, k \in I$, whence $\mathcal{F} = \{J \subseteq I : J_i \subseteq J \text{ for some } i\}$ is a proper filter over I, and by Zorn's Lemma $\mathcal{F}$ can be extended to an ultrafilter $\mathcal{U}$. We claim that $A/\phi_{\mathcal{U}}$ satisfies every sentence in $\mathcal{S}$. This follows from Theorem 5.1 and the observation that for each $\sigma \in \mathcal{S}$,

$$\{j \in I : A_j \text{ satisfies } \sigma\} \supseteq J_{\{\sigma\}} \in \mathcal{U}.$$

Since $\mathcal{K}$ is an elementary class, we have $A/\phi_{\mathcal{U}} \in \mathcal{K}$. But the A_i are all members of $\mathcal{B} - \mathcal{K}$, so this contradicts the assumption that $\mathcal{B} - \mathcal{K}$ is closed under ultraproducts. Therefore $\mathcal{K}$ must be strictly elementary. □

THEOREM 5.3 *Let $\mathcal{K}$ be an elementary class, and suppose $\mathcal{S}$ is some set of sentences such that $\mathcal{K} = \text{Mod}\,\mathcal{S}$. If $\mathcal{K}$ is strictly elementary, then $\mathcal{K} = \text{Mod}\,\mathcal{S}_0$ for some finite set of sentences $\mathcal{S}_0 \subseteq \mathcal{S}$.*

PROOF. Suppose to the contrary that for every finite subset $\mathcal{S}_0$ of $\mathcal{S}$, $\text{Mod}\,\mathcal{S}_0$ properly contains $\mathcal{K}$. As in the proof of the previous theorem we can then construct an ultraproduct $A/\phi_{\mathcal{U}} \in \mathcal{K}$ of algebras A_i not in $\mathcal{K}$. This, however, contradicts the result that the complement of $\mathcal{K}$ is closed under ultraproducts. □

Every identity is a first-order sentence and every variety is an elementary class, so the second result tells us that if a variety is definable by a finite set of first-order sentences, then it is finitely based. The following theorem, from Jónsson [79], uses Theorem 5.2 to give another sufficient condition for a variety to be finitely based.

THEOREM 5.4 *Let $\mathcal{V}$ be a variety of algebras contained in some strictly elementary class $\mathcal{B}$. If there exists an elementary class $\mathcal{C}$ such that $\mathcal{B}_{SI}$ is contained in $\mathcal{C}$ and $\mathcal{V} \cap \mathcal{C}$ is strictly elementary, then $\mathcal{V}$ is finitely based.*

PROOF. Suppose $\mathcal{V}$ is not finitely based. Then Theorem 5.2 implies that $\mathcal{B} - \mathcal{V}$ is not closed under ultraproducts. Hence, for some index set I, there exist $A_i \in \mathcal{B} - \mathcal{V}$ and an ultrafilter $\mathcal{U}$ over I such that the ultraproduct $A/\phi_{\mathcal{U}} \in \mathcal{V}$, where $A = \prod_{i \in I} A_i$.

Each A_i has at least one subdirectly irreducible image A_i' not in $\mathcal{V}$. On the other hand, if we let $A' = \prod_{i \in I} A_i'$ then $A'/\phi_{\mathcal{U}} \in \mathcal{V}$ since it is a homomorphic image of $A/\phi_{\mathcal{U}}$.

$\mathcal{B}$ need not be closed under homomorphic images, so the A_i' are not necessarily in $\mathcal{B}$, but $A'/\phi_{\mathcal{U}} \in \mathcal{V} \subseteq \mathcal{B}$ and $\mathcal{B}$ strictly elementary imply that $\{i \in I : A_i' \in \mathcal{B}\}$ is in $\mathcal{U}$. Therefore, restricting the ultraproduct to this set, we can assume that every $A_i' \in \mathcal{B}_{SI} \subseteq \mathcal{C}$ and, because $\mathcal{C}$ is an elementary class, it follows that $A'/\phi_{\mathcal{U}} \in \mathcal{V} \cap \mathcal{C}$. This contradicts Theorem 5.2 since $\mathcal{V} \cap \mathcal{C}$ is strictly elementary (by assumption), and $A'/\phi_{\mathcal{U}}$ is an ultraproduct of algebras not in $\mathcal{V} \cap \mathcal{C}$. □

Finitely based congruence distributive varieties. Let $\mathcal{V}$ be a congruence distributive variety of algebras (with finitely many operations). By Theorem 1.9 this is equivalent to the existence of $n + 1$ ternary polynomials $t_0, t_1, \ldots, t_n$ such that $\mathcal{V}$ satisfies the following identities:

$$t_0(x, y, z) = x, \qquad t_n(x, y, z) = z, \qquad t_i(x, y, x) = x$$
$$t_i(x, x, z) = t_{i+1}(x, x, z) \qquad \text{for } i \text{ even}$$
$$t_i(x, z, z) = t_{i+1}(x, z, z) \qquad \text{for } i \text{ odd.}$$

In the remainder of this section we let $\mathcal{V}_t$ be the finitely based congruence distributive variety that satisfies these identities. Clearly $\mathcal{V} \subseteq \mathcal{V}_t$.

Translations, boundedness and projective radius. The notion of weak projectivity in lattices and its application to principal congruences can be generalized for an arbitrary algebra A by considering translations of A (i.e. polynomial functions on A with all but one variable fixed in A).

A *0-translation* is any map $f : A \to A$ that is either constant or the identity map. A *1-translation* is a map $f : A \to A$ that is obtained from one of the basic operations of A by fixing all but one variable in A. For our purposes it is convenient to also allow maps that are obtained from one of the polynomials t_i above. Equivalently we could assume that the t_i are among the basic operations of the variety. A *k-translation* is any composition of k 1-translations and a *translation* is a map that is a k-translation for some $k \in \omega$.

For $a, b \in A$ define the relation $\Gamma_k(a, b)$ on A by

$$(c, d) \in \Gamma_k(a, b) \qquad \text{if} \qquad \{c, d\} = \{f(a), f(b)\}$$

for some k-translation f of A. Let $\Gamma(a, b) = \bigcup_{k \in \omega} \Gamma_k(a, b)$. This relation can be used to characterize the principal congruences of A (implicit in Mal'cev [54], see [UA] p.54) as follows:

For $a, b \in A$ we have $(c, d) \in con(a, b)$ if and only if there exists a sequence $c = e_0, e_1, \ldots, e_m = d$ in A such that $(e_i, e_{i+1}) \in \Gamma_k(a, b)$ for $i < m$.

Two pairs $(a, b), (a', b') \in A \times A$ are said to be *k-bounded* if $\Gamma_k(a, b) \cap \Gamma_k(a', b') \neq \mathbf{0}$ and they are *bounded* if $\Gamma(a, b) \cap \Gamma(a', b') \neq \mathbf{0}$. Observe that if A has only finitely many operations, then k-boundedness can be expressed by a first order formula.

The *projective radius* (2-radius in Baker [77]) of an algebra A, written $R(A)$, is the smallest number $k > 0$ such that for all $a, b, a', b' \in A$

$$con(a, b) \cap con(a', b') \neq \mathbf{0} \quad \text{implies} \quad \Gamma_k(a, b) \cap \Gamma_k(a', b') \neq \mathbf{0}$$

(if it exists, else $R(A) = \infty$). For a class $\mathcal{K}$ of algebras, we let $R(\mathcal{K}) = \sup\{R(A) : A \in \mathcal{K}\}$.

The next few lemmas show that under certain conditions a class of finitely subdirectly irreducible algebras (see Section 1.2) is elementary if and only if it has finite projective radius. These results first appeared in a more general form in Baker [77] (using n-radii) but we follow a later presentation due to Jónsson [79].

LEMMA 5.5 *If $A \in \mathcal{V}_t$, $e_0, e_1, \ldots, e_m \in A$ and $e_0 \neq e_m$, then there exists a number $p < m$ such that (e_0, e_m) and (e_p, e_{p+1}) are 1-bounded.*

PROOF. Consider the 1-translations $f_i(x) = t_i(e_0, x, e_m)$, $i \leq n$. Then $f_0(e_j) = e_0$ and $f_n(e_j) = e_m$ for all $i \leq m$, hence there exists a smallest index $q \leq n$ such that the elements

$f_q(e_j)$ are not all equal to e_0. If q is odd, then $f_q(e_0) = f_{q-1}(e_0) = e_0$, so we can choose $p < m$ such that $c = f_q(e_p) = e_0 \neq f_q(e_{p+1}) = d$. It follows that $(c,d) \in \Gamma_1(e_p, e_{p+1})$ and the 1-translation $f(x) = t_q(e_0, e_{p+1}, x)$ shows that $(c,d) \in \Gamma_1(e_0, e_m)$. For even q we have $f_q(e_m) = f_{q-1}(e_m) = e_0$. Choosing $p < m$ such that $c = f_q(e_p) \neq e_0 = f_q(e_{p+1}) = d$, we again see that $(c,d) \in \Gamma_1(e_p, e_{p+1})$, and now the 1-translation $g(x) = t_q(e_0, e_p, x)$ gives $(c,d) \in \Gamma_1(e_0, e_m)$.

In either case $(c,d) \in \Gamma_1(e_0, e_m) \cap \Gamma_1(e_p, e_{p+1})$, which implies that (e_0, e_m) and (e_p, e_{p+1}) are 1-bounded. □

LEMMA 5.6 *For all $A \in \mathcal{V}_t$ and $a, b, a', b' \in A$,*

$$con(a,b) \cap con(a',b') \neq 0 \quad \text{implies} \quad \Gamma(a,b) \cap \Gamma(a',b') \neq 0.$$

PROOF. Suppose $(c,d) \in con(a,b) \cap con(a',b')$ for some $c, d \in A$, $c \neq d$. Since $(c,d) \in con(a,b)$, there exists (by Mal'cev) a sequence $c = e_0, e_1, \ldots, e_m = d$ in A such that $(e_i, e_{i+1}) \in \Gamma(a,b)$ for $i < m$. As before let $f_i(x) = t_i(e_0, x, e_m)$, and choose $p < m$, $q < n$ such that $c' = f_q(e_p) \neq f_q(e_{p+1}) = d'$. By composition of translations $(c', d') \in \Gamma(a,b)$. Also $(c', d') \in con(a',b')$, since $con(a',b')$ identifies e_0 with e_m and hence all elements of the form $t_i(e_0, e_j, e_m)$ with $t_i(e_0, e_j, e_0) = e_0$. Again there exists a sequence $c' = e_0', e_1', \ldots, e_{m'}' = d'$ with $(e_i', e_{i+1}') \in \Gamma(a',b')$ for $i < m'$. From Lemma 5.5 we obtain an index $p < m'$ such that (c', d') and (e_p', e_{p+1}') are 1-bounded and it follows via a composition of translations that (a,b) and (a',b') are bounded. □

Recall from Section 1.2 that an algebra A is finitely subdirectly irreducible if the $0 \in \text{Con}(A)$ is not the meet of finitely many non-0 congruences, and that $\mathcal{V}_{FSI}$ denotes the class of all finitely subdirectly irreducible members of $\mathcal{V}$.

LEMMA 5.7 *Let $\mathcal{C}$ be an elementary subclass of $\mathcal{V}_t$. Then $R(\mathcal{C}_{FSI})$ is finite if and only if $\mathcal{C}_{FSI}$ is elementary.*

PROOF. By assumption algebras in $\mathcal{V}_t$ have only finitely many basic operations, so there exists a first order formula $\phi_k(x, y, x', y')$ such that for all $A \in \mathcal{V}_t$, A satisfies $\phi_k(a, b, a', b')$ if and only if (a,b) and (a',b') are k-bounded. Suppose $R(\mathcal{C}_{FSI}) = k < \infty$. Then an algebra $A \in \mathcal{C}$ is finitely subdirectly irreducible iff it satisfies the sentence σ_k: for all x, y, x', y', $x = y$ or $x' = y'$ or $\phi_k(x, y, x', y')$. Hence $\mathcal{C}_{FSI}$ is elementary. Conversely, suppose $\mathcal{C}_{FSI}$ is an elementary class. Lemma 5.6 implies that $A \in \mathcal{V}_t - \mathcal{C}_{FSI}$ iff A satisfies the negation of σ_k for each $k \in \omega$. So $\mathcal{V}_t - \mathcal{C}_{FSI}$ is also elementary and hence (by Theorem 5.2) strictly elementary, i.e. it is defined by finitely many of the $\neg \sigma_k$. Since $\neg \sigma_{k+1}$ implies $\neg \sigma_k$, we in fact have $A \in \mathcal{V}_t - \mathcal{C}_{FSI}$ iff A satisfies $\neg \sigma_k$ for just one particular k (the largest). It follows that all algebras in $\mathcal{C}_{FSI}$ satisfy σ_k, whence $R(\mathcal{C}_{FSI}) = k$. □

LEMMA 5.8 *If $R(\mathcal{V}_{FSI}) = k < \infty$, then $R(\mathcal{V}) \leq k + 2$.*

PROOF. Let $R(\mathcal{V}_{FSI}) = k < \infty$ and suppose $(c,d) \in con(a_0, b_0) \cap con(a_1, b_1)$ for some $A \in \mathcal{V}$ and $a_0, b_0, a_1, b_1, c, d \in A$, $c \neq d$. Then there exists a subdirectly irreducible epimorphic image A' of A with $c' \neq d'$ and hence $a_0' \neq b_0'$ and $a_1' \neq b_1'$ (primes denote images in A'). By assumption (a_0', b_0') and (a_1', b_1') are k-bounded, i.e. for some distinct $u, v \in A'$, $(u,v) \in \Gamma_k(a_0', b_0') \cap \Gamma_k(a_1', b_1')$. For $i = 0, 1$ choose $u_i, v_i \in A$ such that $(u_i, v_i) \in \Gamma_k(a_i, b_i)$ and $u_i' = u$, $v_i' = v$. Such elements exist since if f' is a k-translation in A' with $f'(a_i') = u$

and $f'(b_i') = v$, then we can construct a corresponding k-translation f in A by replacing each fixed element of A' by one of its preimages in A and we let $u_i = f(a_i)$ and $v_i = f(b_i)$. Now choose $j < n$ such that $u^* = t_j(u_0, u_1, v_0) \neq t_j(u_0, v_1, v_0) = v^*$. This is possible since in A', $t_j(u, u, v)$ and $t_j(u, v, v)$ must be distinct for some $j < n$, else

$$u = t_0(u, u, v) = t_1(u, u, v) = t_1(u, v, v) = t_2(u, v, v) = \ldots = t_n(u, v, v) = v.$$

The 1-translations $t_j(u_0, x, v_0)$, $t_j(u_0, u_1, x)$ and $t_j(u_0, v_1, x)$ now show that $(u^*, v^*) \in \Gamma_1(u_1, v_1)$ and $(u^*, u_0), (u_0, v^*) \in \Gamma_1(u_0, v_0)$. Lemma 5.5 applied to the sequence u^*, u_0, v^* implies that either (u^*, v^*) and (u^*, u_0) are 1-bounded or (u^*, v^*) and (u_0, v^*) are 1-bounded. In either case (u_0, v_0) and (u_1, v_1) are 2-bounded and therefore (a_0, b_0) and (a_1, b_1) are $(k + 2)$-bounded.   □

With the help of these four lemmas and Theorem 5.4, we can now prove the following result:

THEOREM 5.9 (Jónsson [**79**]). *If V is a congruence distributive variety of algebras and V_{FSI} is strictly elementary, then V is finitely based.*

PROOF. By Lemma 5.7 $R(V_{FSI}) = k$ for some $k \in \omega$, and by the above Lemma $R(V) = k + 2$. Let B be the class of all $A \in V_t$ with $R(A) \leq k + 2$. Since the condition $R(A) \leq k + 2$ can be expressed by a first-order formula, and since V_t is strictly elementary, so is B. Clearly $R(B_{FSI}) \leq k + 2$, hence Lemma 5.7 implies that B_{FSI} is elementary. By assumption $V \cap B_{FSI} = V_{FSI}$ is strictly elementary, so applying Theorem 5.4 with $C = B_{FSI}$, we conclude that V is finitely based.   □

Assuming that V is a finitely generated congruence distributive variety, Corollary 1.7 implies that up to isomorphism V_{FSI} is a finite set of finite algebras. Since such a collection is always strictly elementary, one obtains Baker's result from the preceding theorem:

THEOREM 5.10 (Baker [**77**]). *If V is a finitely generated congruence distributive variety of algebras then V is finitely based.*

5.3 Joins of finitely based varieties

In this section we first give an example which shows that the join of two finitely based modular varieties need not be finitely based.

LEMMA 5.11 (Baker [**77'**]). *There exist finitely based modular varieties V and V' such that the complement of $V + V'$ is not closed under ultraproducts.*

PROOF. Let M be the modular lattice of Figure 5.1 (i) and let $N(M)$ be the class of all lattices that do not contain a subset order-isomorphic to M regarded as a partially ordered set). By Lemma 3.10 (ii) $M \notin N(M)$, so there exists an identity $\varepsilon \in \operatorname{Id} N(M)$ that does not hold in M. Let V be the variety of modular lattices that satisfy ε (i.e. $V = \operatorname{Mod}\{\varepsilon, \varepsilon_m\}$, where ε_m is the modular identity) and let V' be the variety of all the dual lattices of members in V. Since the modular variety is self-dual, V' is defined by ε_m and the dual identity ε' of ε. Hence V and V' are both finitely based.

Let K_n be the lattice of Figure 5.1 (ii).

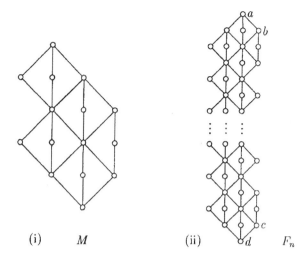

Figure 5.1

We claim that, for each $n \in \omega$, $K_n \notin V + V'$. Note that K_n is subdirectly irreducible (in fact simple), and since K_n contains a copy of M and its dual as sublattices, both ε and ε' fail in K_n. Hence $K_n \notin V \cup V'$, and the claim follows from Theorem 2.3 (i).

Now let $K = \prod_{n \in \omega} K_n$ and choose any nonprincipal ultrafilter $\mathcal{U}$ over ω. We show that the ultraproduct $\overline{K} = K/\phi_{\mathcal{U}}$ is in $V + V'$. Notice that an order-isomorphic copy of M is situated only at the bottom of each K_n. This fact can be expressed as a first-order sentence and, by Theorem 5.1, also holds in $\overline{K}$. Similarly, the dual of M can only be situated at the top of $\overline{K}$. The local structure of the middle portion of K_n can also be described by a first order sentence, whence $\overline{K}$ looks like an infinite version of K_n. Interpreting Figure 5.1 (ii) as a diagram of $\overline{K}$, we see that M is not order-isomorphic to any subset of $\overline{K}/\mathrm{con}(c, d)$ since $\mathrm{con}(c, d)$ collapses the only copy of M in $\overline{K}$. Consequently $\overline{K}/\mathrm{con}(c, d) \in V$ and by a dual argument $\overline{K}/\mathrm{con}(a, b) \in V'$. Observe that $\overline{K}$ is not a simple lattice since principal congruences can only identify quotients reachable by finite sequences of transpositions (Theorem 1.11). In fact $\mathrm{con}(a, b) \cap \mathrm{con}(c, d) = 0$, and hence $\overline{K}$ can be embedded in $K/\mathrm{con}(a, b) \times K/\mathrm{con}(c, d)$. Therefore $\overline{K} \in V + V'$.   □

Together with Theorem 5.2 and Theorem 5.3, the above lemma implies:

THEOREM 5.12 (Baker [77']). *The join of two finitely based (modular) varieties need not be finitely based.*

In view of this theorem it is natural to look for sufficient conditions under which the join of two finitely based varieties is finitely based. In what follows, we shall assume that V_c is a congruence distributive variety, and that V and V' are two subvarieties defined relative to V_c by the identities $p = q$ and $p' = q'$ respectively.

By an elementary result of lattice theory, any finite set of lattice identities is equivalent to a single identity (relative to the class of all lattices, [GLT] p.28). Moreover Baker [74] showed that this result extends to congruence distributive varieties in general. Consequently, the above condition on the varieties V and V' is equivalent to them being finitely based relative to V_c. The next two lemmas are due to Jónsson [74], though the second one has been generalized to congruence distributive varieties.

If p is an n-ary polynomial function (= word or term with at most n variables) on an algebra A and $u_1, \ldots, u_n \in A$ then we will abbreviate $p(u_1, \ldots, u_n)$ by $p(u)$, $u \in A^\omega$, thereby assuming that only the first n components of u are used to evaluate p.

LEMMA 5.13 *An algebra $A \in V_c$ belongs to $V + V'$ if and only if for all $u, v \in A^\omega$*

$$(*) \qquad con(p(u), q(u)) \cap con(p'(v), q'(v)) = 0.$$

PROOF. Let $\theta = \sum\{con(p(u), q(u)) : u \in A^\omega\}$ and $\theta' = \sum\{con(p'(u), q'(u)) : u \in A^\omega\}$. By the (infinite) distributivity of $\mathrm{Con}(A)$ we have that $(*)$ holds if and only if $\theta \cap \theta' = 0$. This in turn is equivalent to A being a subdirect product of A/θ and A/θ'. Since $A/\theta \in V$ and $A/\theta' \in V'$, it follows that $A \in V + V'$. On the other hand Jónsson's Lemma implies that any $A \in V + V'$ can be written as a subdirect product of two algebras A/ϕ and A/ϕ'. Notice that θ and θ' above are the smallest congruences on A for which $A/\theta \in V$ and $A/\theta' \in V'$, hence $\theta \subseteq \phi$ and $\theta' \subseteq \phi'$. Since $\phi \cap \phi' = 0$ we conclude that $\theta \cap \theta' = 0$.   □

Recall the notion of k-boundedness defined in the previous section. It is an elementary property, so we can construct a first-order sentence σ_k such that an algebra $A \in V_c$ satisfies σ_k if and only if for $u, v \in A^\omega$ $(p(u), q(u))$ and $(p'(v), q'(v))$ are not k-bounded.

LEMMA 5.14 *$V + V'$ is finitely based relative to V_c if and only if the following property holds for some positive integer n:*

P(n): *For any $A \in V_c$, if A satisfies σ_n, then A satisfies σ_k for all $k > 1$.*

PROOF. Firstly, we claim that, relative to V_c, the variety $V + V'$ is defined by the set of sentences $S = \{\sigma_1, \sigma_2, \sigma_3, \ldots\}$. Indeed, by Lemma 5.6 we have that

$$con(p(u), q(u)) \cap con(p'(v), q'(v)) = 0$$
$$\text{if and only if}$$
$$\Gamma_k(p(u), q(u)) \cap \Gamma_k(p'(v), q'(v)) = 0$$

for all $k > 0$. Hence by Lemma 5.13 an algebra $A \in V_c$ belongs to $V + V'$ if and only if A satisfies σ_k for all $k > 0$. We can now make use of Theorem 5.3 to conclude that $V + V'$ will have a finite basis relative to V_c if and only if it is defined, relative to V_c, by a finite subset of S, or equivalently by a single sentence σ_k, since σ_k implies σ_m for all $m < k$. If P(n) holds, then clearly $V + V'$ is defined relative to V_c by the sentence σ_n. On the other hand, if P(n) fails, then there must exist an algebra $A \in V_c$ such that A satisfies σ_n but fails σ_m for some $m > n$. If this is true for any positive integer n, then $V + V'$ cannot be finitely based relative to V_c.   □

Although P(n) characterizes all those pairs of finitely based congruence distributive subvarieties whose join is finitely based, it is not a property that is easily verified. Fortunately, for lattice varieties, k-boundedness can be expressed in terms of weak projectivities. More precisely, if we exclude the use of the polynomials t_i in the definition of a k-translation then a k-translation from one quotient of a lattice to another is nothing else but a sequence of k weak transpositions. Two quotients a/b and a'/b' are then said to be k-bounded if they both project weakly onto some nontrivial quotient c/d in less than or equal to k steps. Furthermore, if $p = q$ is a lattice identity then we can assume that the inclusion $p \le q$ holds in any lattice (if not, replace $p = q$ by the equivalent identity $pq = p + q$) and the sentence σ_k can be rephrased as:

$L \in \mathcal{V}_c$ satisfies σ_k if and only if, for all $u, v \in L^\omega$, the quotients $q(u)/p(u)$ and $q'(v)/p'(v)$ do not both project weakly onto a common nontrivial quotient in k (or less) steps.

The following is a slightly sharpened version (for lattices) of Lemma 5.8.

LEMMA 5.15 Let $\overline{L}$ be a homomorphic image of L and let x/y be a prime quotient in L. For any quotient a/b of L, if $\overline{a}/\overline{b}$ projects weakly onto $\overline{x}/\overline{y}$ in n steps, then a/b projects weakly onto x/y in $n + 1$ steps if $n > 0$, and in two steps if $n = 0$.

PROOF. Suppose $\overline{a}/\overline{b}$ projects onto $\overline{x}/\overline{y}$ in 0 steps, i.e. $\overline{a} = \overline{x}$ and $\overline{b} = \overline{y}$. Then

$$a/b \nearrow_w a + y/b + y \searrow_w (a + y)x/(b + y)x = x/y$$
$$\text{and} \, a/b \searrow_w ax/bx \nearrow_w ax + y/bx + y = x/y,$$

since $y \le (b + y)x < (a + y)x \le x$ and $y \le bx + y < ax + y \le x$. Now suppose that $\overline{a}/\overline{b}$ projects weakly onto $\overline{x}/\overline{y}$ in $n > 0$ steps. Since the other cases can be treated similarly, we may assume that $\overline{a}/\overline{b} \nearrow_w \overline{a}_1/\overline{b}_1 \searrow_w \cdots \searrow_w \overline{a}_{n-1}/\overline{b}_{n-1} \nearrow_w \overline{x}/\overline{y}$ for some $b_i, a_i \in L$, $i = 1, \dots, n - 1$. In this case $\overline{b} \le \overline{b}_1$ implies that there exists $b'_1 \in L$ with $\overline{b}'_1 = \overline{b}_1$ and $b \le b'_1$. Letting $a'_1 = b'_1 + a$ we have $\overline{a}'_1 = \overline{a}_1$ and $a/b \nearrow_w a'_1/b'_1$. Next, there exists $a'_2 \in L$ such that $\overline{a}'_2 = \overline{a}_2$ and $a'_2 \le a'_1$. Letting $b'_2 = a'_2 b'_1$ we have $\overline{b}'_2 = \overline{b}_2$ and $a'_1/b'_1 \searrow_w a'_2/b'_2$. Repeating this process we get

$$a/b \nearrow_w a'_1/b'_1 \searrow_w a'_2/b'_2 \nearrow_w \cdots \searrow_w a'_{n-1}/b'_{n-1} \nearrow_w x'/y',$$

where $\overline{x}' = \overline{x}$ and $\overline{y}' = \overline{y}$. By the first argument $x'/y' \nearrow_w x' + x/y' + x \searrow_w x/y$, so a/b projects weakly onto x/y in $n + 2$ steps. One of the steps (x'/y') can still be eliminated, hence the result follows. □

Given a variety $\mathcal{V}$, we denote by $(\mathcal{V})^n$ the variety that is defined by the identities of $\mathcal{V}$ which have n or less variables for some positive integer n. Clearly $\mathcal{V} \subseteq (\mathcal{V})^n$ and $F_{\mathcal{V}}(m) = F_{(\mathcal{V})^n}(m)$ for any $m \le n$. Another nice consequence of this definition is the following lemma, which appears in Jónsson [74].

LEMMA 5.16 If $F_{\mathcal{V}}(n)$ and $F_{\mathcal{V}'}(n)$ are finite for some lattice varieties $\mathcal{V}$ and $\mathcal{V}'$, then $(\mathcal{V} + \mathcal{V}')^n$ is finitely based.

PROOF. In general, if $F_{\mathcal{V}}(n)$ is finite, then $(\mathcal{V})^n$ is finitely based. Now $F_{\mathcal{V}+\mathcal{V}'}(n)$ is a subdirect product of the two finite lattices $F_{\mathcal{V}}(n)$ and $F_{\mathcal{V}'}(n)$, hence finite, and so the result follows. □

We can now give sufficient conditions for the join of two finitely based lattice varieties to be finitely based. This result appeared in Lee [85'] and is a generalization of a result of Jónsson [74].

THEOREM 5.17 If $\mathcal{V}$ and $\mathcal{V}'$ are finitely based lattice varieties with $\mathcal{V} \cap \mathcal{V}' = \mathcal{W}$ and $R(\mathcal{W}_{SI}) = r < \infty$ and if $F_{\mathcal{V}}(r + 3)$ and $F_{\mathcal{V}'}(r + 3)$ are finite, then $\mathcal{V} + \mathcal{V}'$ is finitely based.

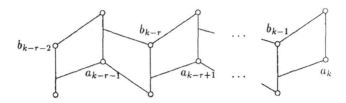

Figure 5.2

PROOF. We can assume that V and V' are defined by the identities $p = q$ and $p' = q'$ respectively, relative to the variety of all lattices, and that the inequalities $p \leq q$ and $p' \leq q'$ hold in any lattice. Let $\mathcal{K}$ and $\mathcal{K}'$ be the classes of $(r + 3)$-generated subdirectly irreducible lattices in V and V' respectively, and define $h = \max(R(\mathcal{K}), R(\mathcal{K}))$. We only consider $h > 0$ since if $h = 0$, then $V, V' \subseteq \mathcal{D}$, the variety of distributive lattices, in which case the theorem holds trivially. Let $V_c = (V + V')^{r+3}$, then V_c is finitely based by Lemma 5.16. If we can show that the condition $P(n)$ in Lemma 5.14 holds for some n, then $V + V'$ will be finitely based relative to V_c and hence relative to the variety of all lattices.

So let $L \in V_c$ and suppose that for some $u, u' \in L^\omega$ the quotients $q(u)/p(u)$ and $q'(u')/p'(u')$ are bounded, that is they project weakly onto a common quotient c/d of L in m and m' steps respectively. Property $P(n)$ demands that $m, m' \leq n$ for some fixed integer n. Take $n = \max(2h + 5, h + r + 5)$ and assume that u, u', c, d have been chosen so as to minimize the number $m + m'$. We will show that if $m > n$ then there is another choice for u, u', c, d such that the corresponding combined number of steps in the weak projectivities is strictly less than $m + m'$. This contradiction, together with the same argument for m', proves the theorem.

By assumption $q(u)/p(u) = a_0/b_0 \sim_w a_1/b_1 \sim_w \ldots \sim_w a_m/b_m = c/d$ for some quotients a_i/b_i in L which transpose weakly alternatingly up and down onto a_{i+1}/b_{i+1} $(i = 0, 1, \ldots, m - 1)$. Since $m > \max(2h + 5, h + r + 5)$, we can always find an integer k such that $\max(h + 2, r + 2) < k < m - h - 2$. Consider the $r + 3$ quotients up to and including a_k/b_k in the above sequence. Since the other cases can be treated similarly, we may assume that

$$a_{k-r-2}/b_{k-r-2} \nearrow_w a_{k-r-1}/b_{k-r-1} \searrow_w \ldots \nearrow_w a_k/b_k.$$

Let L_0 be the sublattice of L generated by the $r + 3$ elements

$$a_{k-r-2}, b_{k-r-1}, a_{k-r}, \ldots, a_{k-1}, b_k.$$

Notice that $L_0 \in V + V'$, and L_0 is a finite lattice because $F_{V+V'}(r+3) = F_{V_c}(r+3)$ is a subdirect product of $F_V(r+3)$ and $F_{V'}(r+3)$, and is therefore finite. $a_k/b_k \ (= a_{k-1}+b_k/b_k)$ can be divided into (finitely many) prime quotients in L_0 and at least one of these prime quotients, say x/y, must project weakly onto a nontrivial subquotient of c/d. Let $\overline{L}_0$ be the unique subdirectly irreducible quotient lattice of L_0 in which $\overline{x}/\overline{y}$ is a critical quotient.

Then $\overline{L}_0 \in \mathcal{V} + \mathcal{V}'$, and hence Theorem 2.3 (i) implies $\overline{L}_0 \in \mathcal{V} \cup \mathcal{V}'$. We examine each of the three cases that arise:

Case 1: $\overline{L}_0 \in \mathcal{V}$ and $\overline{L}_0 \notin \mathcal{V}'$. Since $\overline{L}_0 \notin \mathcal{V}'$, there exists $v \in L_0{}^\omega$ such that $p'(\overline{v}) < q'(\overline{v})$. $\overline{L}_0 \in \mathcal{V}$ implies $R(\overline{L}_0) \le h$. Also $p'(\overline{v}) = \overline{p'(v)}$ and $q'(\overline{v}) = \overline{q'(v)}$. So by Lemma 5.15 $q'(v)/p'(v)$ projects weakly onto x/y in $h + 1$ steps. Now $q(u)/p(u)$ projects weakly onto c/d in k steps, hence onto x/y in $k + 1$ steps. But $h + 1 + k + 1 < m \le m + m'$, so this contradicts the minimality of $m + m'$.

Case 2: $\overline{L}_0 \notin \mathcal{V}$ and $\overline{L}_0 \in \mathcal{V}'$. Since $\overline{L}_0 \notin \mathcal{V}$, $p(v) < q(v)$ for some $v \in L_0{}^\omega$. As above, since $\overline{L}_0 \in \mathcal{V}'$, $R(L_0) \le h$, and hence $q(v)/p(v)$ projects weakly onto x/y in $h + 1$ steps and from there onto a nontrivial subquotient c'/d' of c/d in $m - k$ steps. By the choice of k we have $h + 1 + m - k < m - 1$. Also $q'(u')/p'(u')$ projects weakly onto c/d in $m' + 1$ steps so again we get a contradiction.

Case 3: $\overline{L}_0 \in \mathcal{V} \cap \mathcal{V}' = \mathcal{W}$. First suppose that $r > 0$, hence $\mathcal{W} \ne \mathcal{D}$. $R(\mathcal{W}_{SI}) = r$ implies $\overline{a}_{k-r-2}/\overline{b}_{k-r-2}$ projects weakly onto $\overline{x}/\overline{y}$ in r steps, so by Lemma 5.15 a_{k-r-2}/b_{k-r-2} projects weakly onto x/y in $r + 1$ steps. Now either

$$a_{k-r-2}/b_{k-r-2} \searrow_w a'_{k-r-1}/b'_{k-r-1} \nearrow_w \cdots \searrow_w a'_{k-2}/b'_{k-2} \nearrow_w x/y \quad \text{or}$$
$$a_{k-r-2}/b_{k-r-2} \nearrow_w a'_{k-r-1}/b'_{k-r-1} \searrow_w \cdots \nearrow_w a'_{k-2}/b'_{k-2} \searrow_w x/y$$

for some quotients $a'_{k-r-1}/b'_{k-r-1}, \ldots, a'_{k-2}/b'_{k-2}$ in L. Since

$$a_{k-r-3}/b_{k-r-3} \searrow_w a_{k-r-2}/b_{k-r-2} \quad \text{and} \quad a_k/b_k \searrow_w a_{k+1}/b_{k+1},$$

we have that $q(u)/p(u)$ projects weakly onto x/y in $k - 2$ steps and hence onto a nontrivial subquotient c'/d' of c/d in $m - 2$ steps. As before $q(u')/p(u')$ projects weakly onto c'/d' in $m' + 1$ steps which again contradicts the minimality of $m + m'$.

Now suppose that $r = 1$, which implies $\mathcal{W} = \mathcal{D}$ and $\overline{L}_0 = \mathbf{2}$. Hence in L_0 we have

$$a_{k-2}/b_{k-2} \nearrow a_{k-2} + y/b_{k-2} + y \searrow x/y \quad \text{and}$$

$$a_{k-2}/b_{k-2} \searrow a_{k-2}x/b_{k-2}x \nearrow x/y.$$

It follows that we can shorten the sequence of weak projectivities from $q(u)/p(u)$ onto a nontrivial subquotient c'/d' of c/d to $m - 2$ steps. Again $q'(u')/p'(u')$ projects weakly onto c'/d' in $m' + 1$ steps, giving rise to another contradiction. This concludes the proof. $\square$

We end this section with a theorem that summarizes some conditions under which the join of two finitely based varieties is known to be finitely based. Parts (i) and (ii) are from Lee [85'], and they follows easily from the preceding theorem. Part (iii) is due to Jónsson and the remaining results are from Kang [87].

THEOREM 5.18 *Let $\mathcal{V}$ and $\mathcal{V}'$ be two finitely based lattice varieties. If one of the following conditions holds then $\mathcal{V} + \mathcal{V}'$ is finitely based:*

(i) *$\mathcal{V}$ is modular and $\mathcal{V}'$ is generated by a finite lattice that excludes M_3.*

(ii) *$\mathcal{V}$ and $\mathcal{V}'$ are locally finite and $R(\mathcal{V} \cap \mathcal{V}')$ is finite.*

(iii) *$\mathcal{V}$ and $\mathcal{V}'$ are modular and $\mathcal{V}'$ is generated by a lattice of finite length.*

(iv) *$\mathcal{V}$ is modular and $\mathcal{V}'$ is generated by a lattice with finite projective radius.*

(v) $\mathcal{V} \cap \mathcal{V}' = \mathcal{D}$, the distributive variety.

Lee [85'] also showed that any almost distributive (see Section 4.3) subdirectly irreducible lattice has a projective radius of at most 3. Since any almost distributive variety
is locally finite, it follows from Theorem 5.17 that the join of two finitely based almost
distributive varieties is again finitely based.

5.4   Equational Bases for some Varieties

A variety $\mathcal{V}$ is usually specified in one of two ways: either by a set $\mathcal{E}$ of identities that
determine $\mathcal{V}$ (i.e. $\mathcal{V} = \text{Mod}\,\mathcal{E}$) or by a class $\mathcal{K}$ of algebras that generate $\mathcal{V}$ (i.e. $\mathcal{V} = \mathcal{K}^{\mathcal{V}}$).
In the first case $\mathcal{E}$ is of course an equational basis for $\mathcal{V}$, so here we are interested in the
second case.

A lattice inclusion or inequality of the form $p \leq q$ will also be referred to as a lattice
identity, since it is equivalent to the identity $p = pq$.

Theorem 3.32 shows that the variety $\mathcal{M}_\omega$, generated by all lattices of length 2, has an
equational basis consisting of one identity ε: $x_0(x_1 + x_2 x_3)(x_2 + x_3) \leq x_1 + x_0 x_2 + x_0 x_3$.
Jónsson [68] observed that if one adds to this the identity

$$\varepsilon_n: \quad x_0 \prod_{1 \leq i,j \leq n} (x_i + x_j) \leq \sum_{1 \leq i \leq n} x_0 x_i$$

then one obtains an equational basis for $\mathcal{M}_n = \{M_n\}^{\mathcal{V}}$ ($3 \leq n \in \omega$). To see this, note that
ε_n holds in a lattice of length 2 whenever two of the variables $x_0, x_1, \ldots, x_n$ are assigned
to the same element or one of them is assigned to 0 or 1, but fails when they are assigned
to $n + 1$ distinct atoms. Therefore ε_n holds in $\mathcal{M}_n$ and fails in M_{n+1}.

For $\mathcal{M}_3$ this basis may be simplified even further by observing that ε_3 implies ε, hence

$$\mathcal{M}_3 = \text{Mod}\{x_0(x_1 + x_2)(x_2 + x_3)(x_3 + x_1) \leq x_0 x_1 + x_0 x_2 + x_0 x_3.\}$$

An equational basis for $\mathcal{N}$ was found by McKenzie [72]. It is given by the identities

$$\eta_1: \quad x(y + u)(y + v) \leq x(y + uv) + xu + xv$$
$$\eta_2: \quad x(y + u(x + v)) = x(y + ux) + x(xy + uv)$$

McKenzie shows that η_1 and η_2 hold in any lattice of width ≤ 2, whence $\mathcal{N} \subseteq \text{Mod}\{\eta_1, \eta_2\}$,
and then proves by direct computation that any identity which holds in $\mathcal{N}$ is implied by
η_1 and η_2. In view of Theorem 4.19 the second part may now also be verified by checking
that either η_1 or η_2 fail in each of the lattices $M_3, L_1, L_2, \ldots, L_{15}$ (see Figure 2.2).

Theorem 5.17 implies that the variety $\mathcal{M}^+ = \mathcal{M} + \mathcal{N}$ is finitely based ($\mathcal{M}$ is the variety
of all modular lattices). Note that since $\mathcal{N}$ is the only nonmodular variety that covers
the distributive variety, $\mathcal{M}^+$ is the unique cover of $\mathcal{M}$. Jónsson [77] derives the following
equational basis for $\mathcal{M}^+$ consisting of 8 identities:

(i) $((x + c)y + z)(x + z + a) = (x + a)y + z$

(ii) $(x + c)y \leq x + (y + a)c$

(iii) $((t + x)y + a)c = ((ct + x)y + a)c + ((bt + x)y + a)c$

(iv) $((ct + x)y + a)c = (((ct + x)c + a + xy)y + a)c$

(v) $((bt + x)y + a)c = ((bt + a)c + xy)((b + x)y + a)c$

and the duals of (iii), (iv) and (v), where $a = pq + pr$, $b = q$ and $c = p(q + rq)$.

(Note that (ii) is the identity (AD) which forms part of the equational basis for the variety of all almost distributive lattices in Section 4.3.)

Varieties generated by lattices of bounded length or width. Let $\mathcal{V}_n^m$ be the lattice variety generated by all lattices of length at most m and width at most n $(1 \leq m, n \leq \infty)$ and recall from Section 3.4 the varieties $\mathcal{M}_n^m$ which are defined similarly for modular lattices. For $m, n < \infty$ all these varieties are finitely generated, hence finitely based (Theorem 5.10), and it would be interesting to find a finite equational basis for each of them. Apart from several trivial cases, and the case $\mathcal{M}_n^2 = \mathcal{M}_n$, not much is known about these varieties.

Nelson [68] showed that $\mathcal{V}_2^\infty = \mathcal{N}$ $(= \mathcal{V}_2^n$ for $n \geq 3)$. With the help of Theorem 4.19, this follows from the observation that each of the lattices $M_3, L_1, L_2, \ldots, L_{15}$ has width ≥ 3.

Baker [77] proves that $\mathcal{V}_4^\infty$ and $\mathcal{M}_5^\infty$ are not finitely based, and the same holds for $\mathcal{V}_n^\infty$, $\mathcal{M}_n^\infty$ $n \geq 5$. The proofs are similar to the proof of Lemma 5.11.

As mentioned at the end of Chapter 3 $\mathcal{M}_3^\infty = \mathcal{M}_3$, and by a result of Freese [77] $\mathcal{M}_4^\infty$ is finitely based. Whether $\mathcal{V}_3^\infty$ is finitely based is apparently still an unresolved question.

Chapter 6

Amalgamation in Lattice Varieties

6.1 Introduction

The word amalgamation generally refers to a process of combining or merging certain structures which have something in common, to form a larger or more complicated structure which incorporates all the individual features of its substructures. In the study of varieties, amalgamation, of course, has a very specific meaning, which is defined in the following section. This leads to the formulation of the amalgamation property, which has been of interest for quite some time in several related areas of mathematics such as the theory of field extensions, universal algebra, model theory and category theory.

Amalgamations of groups were originally considered by Schreier [27] in the form of free products with amalgamated subgroup. Implicit in his work, and in the subsequent investigations of B. H. Neumann [54] and H. Neumann [67], is the result that the variety of all groups has the amalgamation property. The first definition of this property in a universal algebra setting can be found in Fraïssé [54]. The strong amalgamation property appears in Jónsson [56] and [60] among a list of properties used for the construction of universal (and homogeneous) models of various first order theories, including lattice theory. One of the results in the [56] paper is that the variety $\mathcal{L}$ of all lattices has the strong amalgamation property. Interesting applications of the amalgamation property to free products of algebras can be found in Jónsson [61], Grätzer and Lakser [71] and [GLT]. The property also plays a role in the theory of algebraic field extensions (Jónsson [62]) and can be related to the solvability of algebraic equations (Hule [76],[78],[79]).

However, it soon became clear that not many of the better known varieties of algebras satisfy the amalgamation property. Counterexamples showing that it fails in the variety of all semigroups are given in Kimura [57] and Howie [62], and these can be used to construct counterexamples for the variety of all rings. As far as lattice varieties are concerned, it follows from Pierce [68] that the variety of all distributive lattices does have the amalgamation property, but Grätzer, Jónsson and Lakser [73] showed that this was not true for any nondistributive modular subvariety. Finally Day and Ježek [84] completed the picture for lattice varieties, by showing that the amalgamation property fails in every nondistributive proper subvariety of $\mathcal{L}$. A comprehensive survey of the amalgamation

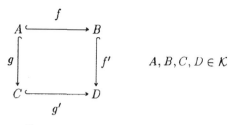

$$A, B, C, D \in \mathcal{K}$$

Figure 6.1

property and related concepts for a wide range of algebras can be found in Kiss, Márki, Pröhle and Tholen [83].

Because of all the negative results, investigations in this field are now focusing on the amalgamation class Amal($\mathcal{K}$) of all amalgamation bases of $\mathcal{K}$, which was first defined in Grätzer and Lakser [71]. A syntactic characterization, and some general facts about the structure of Amal($\mathcal{K}$), $\mathcal{K}$ an elementary class, appear in Yasuhara [74]. Bergman [85] gives sufficient conditions for a member of a residually small variety $\mathcal{V}$ of algebras to be an amalgamation base of $\mathcal{V}$, and Jónsson [90] showed that for finitely generated lattice varieties these conditions are also necessary and, moreover, that it is effectively decidable whether or not a finite lattice is a member of the amalgamation class of such a variety. In Section 6.3 we present some of Bergman's results, and a generalization of Jónsson's results to residually small congruence distributive varieties whose members have one-element subalgebras (due to Jipsen and Rose [89]).

In Grätzer, Jónsson and Lakser [73] it is shown that the two-element chain does not belong to the amalgamation class of any finitely generated nondistributive lattice variety, and that the amalgamation class of the variety of all modular lattices does not contain any nontrivial distributive lattice. On the other hand Berman [81] constructed a nonmodular variety $\mathcal{V}$ such that the two-element chain is a member of Amal($\mathcal{V}$).

Lastly, whenever the amalgamation property fails in some variety $\mathcal{V}$, then Amal($\mathcal{V}$) is a proper subclass of $\mathcal{V}$, and it would be of interest to know what kind of class we are dealing with. In particular, is Amal($\mathcal{V}$) an elementary class (i.e. can membership be characterized by some collection of first order sentences)? Using results of Albert and Burris [88], Bergman [89] showed that the amalgamation class of any finitely generated nondistributive modular variety is not elementary. In contrast Bruyns, Naturman and Rose [a] show that for the variety generated by the pentagon, the amalgamation class is elementary, and is in fact determined by Horn sentences.

6.2 Preliminaries

The amalgamation class of a variety. By a *diagram* in a class $\mathcal{K}$ of algebras we mean a quintuple (A, f, B, g, C) with $A, B, C \in \mathcal{K}$ and $f : A \hookrightarrow B, g : A \hookrightarrow C$ embeddings. An *amalgamation* in $\mathcal{K}$ of such a diagram is a triple (f', g', D) with $D \in \mathcal{V}$ and $f' : B \hookrightarrow D$, $g' : C \hookrightarrow D$ embeddings such that $f'f = g'g$ (see Figure 6.1).

A *strong amalgamation* is an amalgamation with the additional property that

$$f'(B) \cap g'(C) = f'f(A).$$

An algebra $A \in \mathcal{K}$ is called an *amalgamation base* for $\mathcal{K}$ if every diagram (A, f, B, g, C) can be amalgamated in $\mathcal{K}$. The class of all amalgamation bases for $\mathcal{K}$ is called the *amalgamation class* of $\mathcal{K}$, and is denoted by $Amal(\mathcal{K})$. $\mathcal{K}$ is said to have the *(strong) amalgamation property* if every diagram can be (strongly) amalgamated in $\mathcal{K}$. We are interested mainly in the case where $\mathcal{K}$ is a variety.

Some general results about $Amal(\mathcal{V})$. We summarize below some results, the first of which is due to Grätzer, Jónsson and Lakser [73] and the others are from Yasuhara [74].

THEOREM 6.1 *Let $\mathcal{V}$ be a variety of algebras.*

(i) *If $f : A \hookrightarrow A' \in Amal(\mathcal{V})$, and for every $g : A \hookrightarrow C$, f and g can be amalgamated in $\mathcal{V}$, then $A \in Amal(\mathcal{V})$.*

(ii) *Every $A' \in \mathcal{V}$ can be embedded in some $A \in Amal(\mathcal{V})$, with $|A| \leq |A'| + \omega$.*

(iii) *$Amal(\mathcal{V})$ is a proper class.*

(iv) *The complement of $Amal(\mathcal{V})$ is closed under reduced powers.*

(v) *If $A \times A' \in Amal(\mathcal{V})$, and if A' has a one element subalgebra, then $A \in Amal(V)$.*

In general we know very little about the members of $Amal(\mathcal{V})$. Take for example $\mathcal{V} = \mathcal{M}$, the variety of all modular lattices: as yet nobody has been able to construct a nontrivial amalgamation base for $\mathcal{M}$. In fact, we do not even know whether $Amal(\mathcal{M})$ has any finite members except the trivial lattices. As we shall see below, the situation is somewhat better if we restrict ourselves to residually small varieties (defined below).

Essential extensions and absolute retracts. An extension B of an algebra A is said to be *essential* if every nontrivial congruence on B restricts to a nontrivial congruence on A. An embedding $f : A \hookrightarrow B$ is an *essential embedding* if B is an essential extension of $f(A)$. Notice that if A is (a, b)-irreducible (i.e. con(a,b) is the smallest nontrivial congruence on A) and $f : A \hookrightarrow B$ is an essential embedding, then B is $(f(a), f(b))$-irreducible.

LEMMA 6.2 *If $h : A \hookrightarrow B$ is any embedding, then there exists a congruence θ on B such that h followed by the canonical epimorphism from B onto B/θ is an essential embedding of A into B/θ.*

PROOF. By Zorn's Lemma we can choose θ to be maximal with respect to not identifying any two members of $h(A)$. $\square$

An algebra A in a variety $\mathcal{V}$ is an *absolute retract of $\mathcal{V}$* if, for every embedding $f : A \hookrightarrow B$ with $B \in \mathcal{V}$, there exists an epimorphism (retraction) $g : B \twoheadrightarrow A$ such that the composite gf is the identity map on A.

THEOREM 6.3 (Bergman [85]). *Every absolute retract of a variety $\mathcal{V}$ is an amalgamation base of $\mathcal{V}$.*

PROOF. Suppose A is an absolute retract of $\mathcal{V}$ and let (A, f, B, g, C) be a diagram in $\mathcal{V}$. Then there exist epimorphisms h and k such that $fh = id_A = kg$. To amalgamate the diagram, we take $D = B \times C$ and define $f' : B \hookrightarrow D$ by $f'(b) = (b, gh(b))$ and $g' : C \hookrightarrow D$ by $g'(c) = (fk(c), c)$, then $f'f(a) = (f(a), g(a)) = g'g(a)$ for all $a \in A$. $\square$

Recall that V_{SI} denotes the class of all subdirectly irreducible algebras of V and consider the following two subclasses (referred to as the class of all *maximal irreducibles* and all *weakly maximal irreducibles* respectively):

$$V_{MI} = \{M \in V_{SI} : M \text{ has no proper extension in } V_{SI}\}$$
$$V_{WMI} = \{M \in V_{SI} : M \text{ has no proper essential extension in } V_{SI}\}.$$

Clearly $V_{MI} \subseteq V_{WMI}$.

LEMMA 6.4 $M \in V_{WMI}$ *if and only if* $M \in V_{SI}$ *and* M *is an absolute retract in* V.

PROOF. Let $M \in V_{WMI}$ and suppose $f : M \hookrightarrow B \in V$ is an embedding. By Lemma 6.2, f induces an essential embedding $f' : M \hookrightarrow B/\theta$ for some $\theta \in \text{Con}(B)$. Since M has no proper essential extensions, f' must be an isomorphism, so the canonical epimorphism from B to B/θ followed by the inverse of f' is the required retraction. The converse follows from the observation that an absolute retract of V cannot have a proper essential extension in V. $\qquad\square$

Theorem 6.3 together with Lemma 6.4 implies that $V_{WMI} \subseteq \text{Amal}(V)$. Observe also that if V is a finitely generated congruence distributive variety, then V_{SI} is a finite set of finite algebras (Corollary 1.7), and so we can determine the members of V_{MI} by inspection.

Residually small varieties. A variety V is said to be *residually small* if the subdirectly irreducible members of V form, up to isomorphism, a set, or equivalently, if there exists an upper bound on the cardinality of the subdirectly irreducible members of V. For example, any finitely generated congruence distributive variety is residually small.

THEOREM 6.5 (Taylor [72]). *If* V *is a residually small variety, then every member of* V_{SI} *has an essential extension in* V_{WMI}.

PROOF. The union of a chain of essential extensions is again an essential extension, so we can apply Zorn's Lemma to the set V_{SI} (ordered by essential inclusion) to obtain its maximal elements. Clearly these are all the elements of V_{WMI}. $\qquad\sqcap$

In fact Taylor [72] also proved the converse of the above theorem, but we won't make use of this result. Note that if V is a finitely generated congruence distributive variety then every member of V_{SI} has an essential extension in V_{MI}.

Amalgamations constructed from factors. The following lemma is valid in any class of algebras that is closed under products, and makes the problem of amalgamating a diagram somewhat more accessible.

LEMMA 6.6 (Grätzer and Lakser [71]). *A diagram* (A, f, B, g, C) *in a variety* V *can be amalgamated if and only if for all* $u \neq v \in B$ *there exists a* $D \in V$ *and homomorphisms* $f' : B \to D$ *and* $g' : C \to D$ *such that* $f'f = g'g$ *and* $f'(u) \neq f'(v)$, *and the same holds for* C.

PROOF. The condition is clearly necessary. To see that it is sufficient, we need only observe that the diagram can be amalgamated by the product of these D's, generated as u and v run through all distinct pairs of B and C. $\qquad\square$

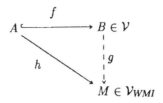

Figure 6.2

6.3   Amal($\mathcal{V}$) for Residually Small Varieties

Property (Q). An algebra A in a variety $\mathcal{V}$ is said to have *property (Q)* if for any embedding $f : A \hookrightarrow B \in \mathcal{V}$ and any homomorphism $h : A \to M \in \mathcal{V}_{WMI}$ there exists a homomorphism $g : B \to M$ such that $h = gf$.

This property was used in Grätzer and Lakser [71], Bergman [85] and Jónsson [90] to characterize amalgamation classes of certain varieties.

THEOREM 6.7 (Bergman [85]). *Let $\mathcal{V}$ be a residually small variety. If $A \in \mathcal{V}$ has property (Q), then $A \in A\mathrm{mal}(\mathcal{V})$.*

PROOF. We use Lemma 6.6. Let (A, f, B, g, C) be any diagram in $\mathcal{V}$ and let $u \neq v \in B$. Choose a maximal congruence θ on B such that θ does not identify u and v. Then $B/\theta \in \mathcal{V}_{SI}$ and hence by Theorem 6.5 B/θ has an essential extension $M \in \mathcal{V}_{WMI}$. Let f' be the canonical homomorphism $B \twoheadrightarrow B/\theta$, but considered as a map into M. Clearly $f'(u) \neq f'(v)$, and since A has property (Q), the homomorphism $f'f : A \to M$ can be extended to a homomorphism $g' : C \to M$ such that $f'f = g'g$. We argue similarly for $u \neq v \in C$, hence Lemma 6.6 implies $A \in A\mathrm{mal}(\mathcal{V})$.   □

We show that, for certain congruence distributive varieties of algebras (including all residually small lattice varieties), the converse of the above theorem also holds. We first prove two simple results.

LEMMA 6.8 (Jipsen and Rose [89]). *Let A and B be algebras in a congruence distributive variety $\mathcal{V}$, $a \in A$ and suppose $\{a\}$ is a subalgebra of A. Let $h_a : B \hookrightarrow A \times B$ be an embedding such that $h_a(b) = (a, b)$ for all $b \in B$. Then the projection $\pi_B : A \times B \to B$ is the only retraction of h_a onto B.*

PROOF. Suppose $g : A \times B \twoheadrightarrow B$ is a retraction, that is gh_a is an identity map on B. By Lemma 1.3 there exist $\theta \in \mathrm{Con}(A)$ and $\phi \in \mathrm{Con}(B)$ such that for $(x, y), (x', y') \in A \times B$

$$(x, y) \ker g \ (x', y') \quad \text{if and only if} \quad x\theta x' \text{ and } y\phi y'.$$

Since gh_a is an identity map on B, ϕ must be a trivial congruence on B. To prove that $g = \pi_B$ it suffices to show that for any $a' \in A$ we have $a\theta a'$. Suppose the contrary. First observe that for $b, b' \in B$ with $b \neq b'$ we always have

$$g(a, b) = b \neq b' = g(a, b').$$

Now if $(a, a') \notin \theta$ for some $a' \in A$, then there exist $b, b' \in B$ such that

$$g(a,b) = b \neq b' = g(a',b).$$

Thus $g(a, b') = g(a', b)$ and so Lemma 1.3 implies $a\theta a'$ and $b\phi b'$, a contradiction. ☐

COUNTEREXAMPLE. The assumption that $h = h_a$ is a one-element subalgebra embedding cannot be dropped. Indeed, let $2 = \{0, 1\}$ be the two-element chain and consider a lattice embedding $h : 2 \hookrightarrow 2 \times 2$ given by $h(0) = (0,0)$ and $h(1) = (1,1)$. Then both projections on 2×2 are retractions onto 2.

COROLLARY 6.9 Let V be congruence distributive, $A, B \in V$, and suppose A has a one-element subalgebra $\{a\}$ and B is an absolute retract in V. If $k : A \times B \hookrightarrow C \in V$ is an embedding, then the projection $\pi_B : A \times B \to B$ can be extended to an epimorphism of C onto B.

PROOF. If $h_a : B \hookrightarrow A \times B$ is an embedding as in Lemma 6.8, then kh_a is an embedding of B into C. Since B is an absolute retract in V there is a retraction p of C onto B. It follows from Lemma 6.8 that $p|A \times B = \pi_B$. ☐

The characterization theorem. The following theorem is a generalization of a result of Jónsson [90]. There the result was proved for finitely generated lattice varieties.

THEOREM 6.10 (Jipsen and Rose [89]). Let V be a residually small congruence distributive variety, $A \in V$ and suppose A has a one-element subalgebra. Then the following conditions are equivalent:

(i) A satisfies property (Q);

(ii) $A \in Amal(V)$;

(iii) Let $h : A \to M \in V_{WMI}$ be a homomorphism and $k : A \hookrightarrow A \times M$ be an embedding given by $k(a) = (a, h(a))$ for all $a \in A$. If $f : A \hookrightarrow B \in V$ is an essential embedding then the diagram $(A, f, B, k, A \times M)$ can be amalgamated in V.

PROOF. (i) implies (ii) by Theorem 6.7, and trivially (ii) implies (iii). Suppose (iii) holds. It follows from Lemma 6.2, that in order to prove (i), we may assume that the embedding $f : A \hookrightarrow B$ is essential. Let $h : A \to M \in V_{WMI}$ be any homomorphism, and define an embedding $k : A \hookrightarrow A \times M$ by $k(a) = (a, h(a))$ for all $a \in A$. Notice that $h = \pi_M k$ where π_M is the projection from $A \times M$ onto M. By (iii) the diagram $(A, f, B, k, A \times M)$ has an amalgamation (C, f', k') in V. It follows from Corollary 6.9 that there is a retraction $p : C \to M$ such that $h = pk'k = pf'f$ (see Figure 6.3). Letting $g = pf'$ we have $h = gf$. ☐

In case V is a finitely generated congruence distributive variety, we have that each $M \in V_{WMI}$ is embedded in some $M' \in V_{MI}$ (= the set of all maximal extensions in the finite set V_{WMI}). Since members of V_{WMI} are absolute retracts, we only have to test property (Q) for all homomorphisms $h : A \to M' \in V_{MI}$ (this is how property (Q) is defined in Jónsson [90]). If $A \in V$ is a finite algebra, then A has only finitely many nonisomorphic essential extensions $B \in V$ and there are only finitely many possibilities for the homomorphisms $h : A \to M' \in V_{MI}$. In each case one can effectively determine if there exists a homomorphism $g : B \to M'$ such that $g|A = h$. Thus we obtain:

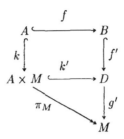

Figure 6.3

COROLLARY 6.11 (Jónsson [**90**]). *Let V be a finitely generated congruence distributive variety. If $A \in V$ is a finite algebra with a one-element subalgebra, then it is effectively decidable whether or not $A \in$ Amal(V).*

Property (P). We conclude this section by stating without proof further interesting results from Bergman [**85**] and Jónsson [**90**].

For an algebra A in a variety V, we let $A^\#$ be the direct product of all algebras A/θ with $\theta \in \text{Con}(A)$ and $A/\theta \in V_{SI} \cap \text{Amal}(V)$, and we let μ_A be the canonical homomorphism of A into $A^\#$.

An algebra A in a variety V is said to have *property (P)* if μ_A is an embedding of A into $A^\#$, and for every homomorphism $g : A \to M \in V_{MI}$ there exists a homomorphism $h : A^\# \to M$ with $h\mu_A = g$.

THEOREM 6.12 (Bergman [**85**]). *For any finitely generated variety of modular lattices and $A \in V$, we have $A \in$ Amal(V) if and only if A is congruence extensile and has property (P).*

Bergman also showed that the above theorem holds for $V = \mathcal{N}$, the smallest non-modular variety (see Jónsson [**90**]), and that the reverse implication holds for all finitely generated lattice varieties, but it is not known whether the same is true for the forward implication.

THEOREM 6.13 (Jónsson [**90**]). *A finite lattice $L \in \mathcal{N}$ belongs to Amal($\mathcal{N}$) if and only if L is a subdirect power of N and L does not have the three element chain as a homomorphic image.*

It is not known whether this theorem is also true for infinite members of $\mathcal{N}$.

6.4   Products of absolute retracts

It is shown in Taylor [**73**] that, in general, the product of absolute retracts is not an absolute retract even if V is a congruence distributive variety. Theorem 6.14 however shows that absolute retracts are preserved under arbitrary products in a congruence distributive varieties, provided that every member of this variety has a one-element subalgebra. The theorem is a generalization of a result of Jónsson [90], which states that if V is a finitely generated lattice variety then any product of members of V_{MI} is an absolute retract in V.

THEOREM 6.14 (Jipsen and Rose [89]). *Let V be a congruence distributive variety and assume that every member of V has a one-element subalgebra. Then every direct product of absolute retracts in V is an absolute retract in V.*

PROOF. Suppose $A = \prod_{i \in I} A_i$ is a direct product of absolute retracts in V, and consider an embedding $f : A \hookrightarrow B \in V$. For $i \in I$, let $\pi_i : A \to A_i$ be a projection. By Corollary 6.9 there is a homomorphism $h_i : B \to A_i$ such that $\pi_i = h_i f$. Consider a homomorphism $h : B \to A$ given by $\pi_i h = h_i$ for each $i \in I$. Then $\pi_i h f = h_i f = \pi_i$ and so $hf : A \to A$ is the identity map. □

6.5 Lattices and the Amalgamation Property

Given the characterization of $\text{Amal}(V)$ (Theorem 6.10), some well known results about the amalgamation classes of finitely generated lattice varieties can be derived easily.

THEOREM 6.15 (Pierce [68]). *The variety D of all distributive lattices has the amalgamation property.*

PROOF. We show that property (Q) is satisfied for any $A \in D$. $D_{SI} = D_{WMI} = D_{MI} = \{2\}$, so let $A, B \in D$ with embedding $f : A \hookrightarrow B$ and homomorphism $h : A \to 2$. If $h(A) = \{0\}$ or $h(A) = \{1\}$ then trivially there exists $g : B \to 2$ such that $h = gf$. On the other hand, if h is onto, then $\ker h$ splits A into an ideal I and a filter F, say. Extend $f(I)$ to the ideal $I' = \{b \in B : b \le a \in f(I)\}$ and $f(F)$ to the filter $F' = \{b \in B : b \ge a \in f(F)\}$. Clearly $I' \cap F' = \emptyset$, hence by Zorn's Lemma and the distributivity of B, I' can be enlarged to a maximal ideal P, which is also disjoint from F'. Define $g : B \to 2$ by $g(b) = 0$ if $b \in P$, and $g(b) = 1$ otherwise. Then one easily checks that g is a homomorphism and that $h = gf$. □

In fact, one can show more generally that if V is any congruence distributive variety which is generated by a finite simple algebra that has no nontrivial subalgebra, then V has the amalgamation property. This result is essentially contained in Day [72].

THEOREM 6.16 *For any nondistributive finitely generated lattice variety V we have $2 \notin \text{Amal}(V)$ and consequently the amalgamation property fails in V.*

PROOF. Since V is nondistributive, M_3 or N is a member of V. Let f be a map that embeds 2 into a prime critical quotient of $L = M_3$ or $L = N$, depending on which lattice is in V. Also, since each $M \in V_{MI}$ is finite, we can define the map $h : 2 \hookrightarrow M$ by $h(0) = 0_M$ and $h(1) = 1_M$. Now it is easy to see that there does not exist a homomorphism $g : L \to M$ such that $h = gf$ (see Figure 6.4). □

Berman [81] showed that there exists a lattice variety V such that $2 \in \text{Amal}(V)$. In fact Berman considers the variety $V = \{L_6^n : n \in \omega\}^V$ (see Figure 2.2) and proves that all of its finitely generated subdirectly irreducible members are amalgamation bases.

L has the strong amalgamation property. Next we would like to prove the result of Jónsson [56], that the variety L of all lattices has the strong amalgamation property. Since L is not residually small, we cannot make use of the results in the previous section.

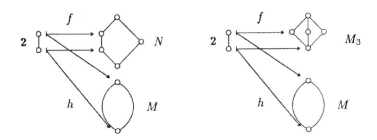

Figure 6.4

We first consider a notion weaker than that of an amalgamation: Let L_1 and L_2 be two lattices with $L' = L_1 \cap L_2$ a sublattice of both L_1 and L_2. A *completion* of L_1 and L_2 is a triple (f_1, f_2, L_3) such that L_3 is any lattice and $f_i : L_i \hookrightarrow L_3$ are embeddings $(i = 1, 2)$ with $f_1 | L' = f_2 | L'$.

Suppose L_1 and L_2 are members of some lattice variety $\mathcal{V}$, and let us denote the inclusion map $L' \hookrightarrow L_i$ by j_i $(i = 1, 2)$. Then clearly (f_1, f_2, L_3) is an amalgamation of the diagram (L', j_1, L_1, j_2, L_2) in $\mathcal{V}$ if and only if $L_3 \in \mathcal{V}$.

How to construct a completion of L_1 and L_2? This can be done in various ways, of which we consider two, namely the ideal completion and the filter completion. We begin by setting $P = L_1 \cup L_2$ and defining a partial order $\leq_P$ on P as follows: on L_1 and L_2, $\leq_P$ agrees with the existing order (which we denote by $\leq_1$ and $\leq_2$ respectively), and if $x \in L_i$, $y \in L_j$ $(\{i, j\} = \{1, 2\})$ then

$$x \leq_P y \qquad \text{if and only if} \qquad \begin{array}{l} x \leq_i z \quad \text{and} \quad z \leq_j y \\ \text{for some} \quad z \in L_1 \cap L_2 \end{array}$$

(equivalently $\leq_P = \leq_1 \cup \leq_2 \cup \leq_1 \circ \leq_2 \cup \leq_2 \circ \leq_1$). It is straightforward to verify that $\leq_P$ is indeed a partial order on P. Define a subset I of P to be a (L_1, L_2)-*ideal* of P if I satisfies

(1) $x \in I$ and $z \leq_P x$ imply $z \in I$ and
(2) $x, y \in I \cap L_i$ implies $x +_i y \in I$ $(i = 1, 2)$.

Let $\mathcal{I}(L_1, L_2)$ be the collection of all (L_1, L_2)-ideals of P together with the empty set. $\mathcal{I}(L_1, L_2)$ is closed under arbitrary intersections, so it forms a complete lattice with $I \cdot J = I \cap J$ and $I + J$ equal to the (L_1, L_2)-ideal generated by $I \cup J$ for any $I, J \in \mathcal{I}(L_1, L_2)$. For each $x \in L_1 \cup L_2$ the principal ideal $(x]$ is in $\mathcal{I}(L_1, L_2)$, so we can define the maps

$$f_i : L_i \hookrightarrow \mathcal{I}(L_1, L_2) \quad \text{by} \quad f_i(x) = (x] \qquad (i = 1, 2).$$

LEMMA 6.17 $(f_1, f_2, \mathcal{I}(L_1, L_2))$ *is a completion (the ideal completion) of* L_1 *and* L_2.

PROOF.   Let $x, y \in L_i$ $(i = 1$ or $2)$. Then $f_i(x +_i y) = (x +_i y] \supseteq (x] + (y]$ since $x, y \leq_P x +_i y$. On the other hand we have $x, y \in (x] + (y]$ so by (2) $x +_i y \in (x] + (y]$,

Figure 6.5

hence $f_i(x +_i y) = f_i(x) + f_i(y)$. Similarly $f_i(x \cdot_i y) = f_i(x)f_i(y)$ and clearly f_1 and f_2 are one-one, with $f_1|L = f_2|L$. □

The notion of a (L_1, L_2)-*filter* and the *filter completion* $(g_1, g_2, \mathcal{F}(L_1, L_2))$ are defined dually. As an easy consequence we now obtain:

THEOREM 6.18 (Jónsson [56]). *$\mathcal{L}$ has the strong amalgamation property.*

PROOF. Let (A, f, B, g, C) be a diagram in $\mathcal{L}$. Since $\mathcal{L}$ is closed under taking isomorphic copies, we may assume that A is a sublattice of B and C, and that $A = B \cap C$ with f and g as the corresponding inclusion maps. Now $\mathcal{I}(B, C) \in \mathcal{L}$, so the ideal completion $(f_1, f_2, \mathcal{I}(B, C))$ is in fact an amalgamation of the diagram. To see that this is a strong amalgamation we need only observe that if $(x] \in f_1(B) \cap f_2(C)$ then $x \in A$. □

Observe that we could not have used the above approach to prove the amalgamation property for the variety $\mathcal{D}$, since the ideal completion of two distributive lattices need not be distributive. Indeed, let $B = M_2(a, b)$ ($= \mathbf{2} \times \mathbf{2}$ generated by a, b) and $C = M_2(a, c)$ with $a + b = a + c$ and $ab = ac$ (see Figure 6.5).

Then $B \cup C = M_3(a, b, c)$ is already a lattice, and therefore a sublattice of $\mathcal{I}(B, C)$. However $B \cup C$ is nondistributive, hence $\mathcal{I}(B, C) \notin \mathcal{D}$. The same holds for any other lattice D in which we might try to amalgamate B and C, so it follows that $\mathcal{D}$ does not have the strong amalgamation property.

6.6 Amalgamation in modular varieties

No nondistributive modular variety has the amalgamation property. In our presentation of this result we follow Grätzer, Jónsson and Lakser [73]. We begin with a technical lemma.

LEMMA 6.19 *Let $\mathcal{V}$ be a variety of algebras that has the amalgamation property, and let $A, B, C, D \in \mathcal{V}$.*

(i) *If D is an extension of C, and f is any embedding of C into D, then there exists an extension $E \in \mathcal{V}$ of D and an embedding $g : D \hookrightarrow E$ such that $g|C = f$.*

(ii) *If B is an extension of A, and α is an automorphism of A, then there exists an extension $\bar{B} \in \mathcal{V}$ of B and an automorphism $\bar{\alpha}$ of $\bar{B}$ such that $\bar{\alpha}|A = \alpha$*

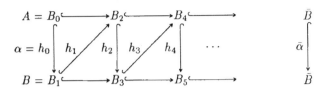

Figure 6.6

PROOF. (i) Let j be the inclusion map $C \hookrightarrow D$ and consider the diagram (C, f, D, j, D) which by assumption has an amalgamation (f', g, E). The result follows if we identify D with its isomorphic image $f'(D)$ in E. To prove (ii), let $B_0 = A$, $B_1 = B$, and consider $h_0 = \alpha$ as an embedding of B_0 into B_1. We now apply part (i), with $C = h_0(B_0)$ and $f = h_0^{-1}$ to obtain an extension B_2 of B_1 and an embedding $h_1 : B_1 \hookrightarrow B_2$ satisfying $h_1|h_0(B_0) = h_0^{-1}$. Repeating this process for $n = 2, 3, \ldots$ we get a sequence of extensions of $A = B_0 \subseteq B_1 \subseteq B_2 \subseteq \ldots$ and a sequence of embeddings $h_n : B_n \hookrightarrow B_{n+1}$ such that $h_{n+1}|h_n(B_n) = h_n^{-1}$ (see Figure 6.6).

We can now define

$$\bar{B} = \bigcup_{n \in \omega} B_n \qquad \text{and} \qquad \bar{\alpha} = \bigcup_{n \in \omega} h_{2n},$$

then clearly $\bar{\alpha}$ is an embedding of $\bar{B}$ into $\bar{B}$. To see that $\bar{\alpha}$ is also onto, choose any $y \in \bar{B}$, then there is an $n \in \omega$ such that $y \in B_{2n+1}$. Put $x = h_{2n+1}(y)$, then $\bar{\alpha}(x) = h_{2n+2}(x) = y$, since h_{2n+2} is an extension of h_{2n+1}^{-1}. Hence $\bar{\alpha}$ is an automorphism of $\bar{B}$, and by construction $\bar{\alpha}|A = \alpha$. $\qquad\qquad\square$

THEOREM 6.20 (Grätzer, Jónsson and Lakser [73]). *Any nondistributive subvariety of the variety $\mathcal{M}$ of all modular lattices does not have the amalgamation property.*

PROOF. Let us assume to the contrary that there exists a variety $\mathcal{V}$ such that $\mathcal{D} \subset \mathcal{V} \subseteq \mathcal{M}$ and $\mathcal{V}$ has the amalgamation property. Under these assumptions we will prove a number of statements about $\mathcal{V}$ that will eventually lead to a contradiction. The proof does require the coordinatization theorem of projective spaces (see Section 3.2). As a simple observation we have that $M_3 \in \mathcal{V}$.

Statement 1: Every member of $\mathcal{V}$ can be embedded in a simple complemented lattice that also belongs to $\mathcal{V}$.

Clearly any lattice in $\mathcal{V}$ can be embedded into some lattice $L \in \mathcal{V}$, which has a least and a greatest element, denoted by 0_L and 1_L respectively (e.g. we can take L to be the ideal lattice). Given $x \in L$, $x \neq 0_L, 1_L$ we can embed the three element chain $3 = \{0 < 1 < 2\}$ into L and into $M_3(a, b, c)$ such that

$$0 \mapsto 0_L, \quad 1 \mapsto x, \quad 2 \mapsto 1_L \qquad \text{in } L$$
$$0 \mapsto ab, \quad 1 \mapsto a, \quad 2 \mapsto a + b \qquad \text{in } M_3(a, b, c).$$

By the amalgamation property, there exists a lattice $L_1 \in \mathcal{V}$ with L as a 0,1-sublattice (i.e. $0_L = 0_{L_1}$ and $1_L = 1_{L_1}$) such that $\{0_L, x, 1_L\}$ is contained in a diamond sublattice of L_1.

Iterating this process for each $x \in L$, $x \neq 0_L, 1_L$ we obtain a lattice $C_1 \in \mathcal{V}$ that contains L as a sublattice, and for each $x \in L$ $x \neq 0_L, 1_L$ there exists $y, z \in C_1$ such that $\{0_L, x, y, z, 1_L\}$ form a diamond sublattice. Repeating this process, we obtain an infinite sequence

$$L = C_0 \subseteq C_1 \subseteq C_2 \subseteq \ldots \subseteq C_n \subseteq \ldots$$

of lattices in $\mathcal{V}$, each with the same 0 and 1 as L and satisfying: for all $n \in \omega$, all $x \in C_n$, if $x \neq 0_L, 1_L$ then x belongs to a diamond $\{0_L, x, y, z, 1_L\}$ in C_{n+1}. Letting $C_\infty = \bigcup_{n \in \omega} C_n$, we have that $C_\infty \in \mathcal{V}$, and clearly C_∞ is complemented. In a complemented lattice, a congruence θ is determined by the ideal $0_L/\theta$, hence if θ is a nontrivial congruence on C_∞, then $x \theta 0_L$ for some $x \in C_\infty$, $x \neq 0_L$. Now $x = 1_L$ implies that θ collapses all of C_∞, and $x < 1_L$ implies that x belongs to a diamond $\{0_L, x, y, z, 1_L\}$ in C_∞, so again θ collapses all of C_∞, since the diamond M_3 is simple. Hence C_∞ is a simple complemented lattice in $\mathcal{V}$ containing L as a sublattice.

Note that since Hall and Dilworth [46] constructed a modular lattice that cannot be embedded in any complemented modular lattice, this statement suffices to conclude that $\mathcal{M}$ does not have the amalgamation property.

Statement 2: For every $L \in \mathcal{V}$ there exists an infinite dimensional nondegenerate projective space P such that L can be embedded in $\mathcal{L}(P)$ and $\mathcal{L}(P) \in \mathcal{V}$.

We may assume that L has a greatest and least element, and that it contains an infinite chain, for if it does not, then we adjoin an infinite chain above the greatest element of L and the resulting lattice is still a member of $\mathcal{V}$. By Statement 1, L can be embedded in a simple complemented lattice $C \in \mathcal{V}$, and by Theorem 3.3, C can be 0,1-embedded in some modular geometric lattice $M \in \mathcal{V}$. M can be represented as a product of modular geometric lattices M_i ($i \in I$) which correspond to nondegenerate projective spaces P_i, in the sense that $M_i \cong \mathcal{L}(P_i)$. Let f_i denote the embedding of C into M followed by the ith projection. Since f_i preserves 0 and 1, it cannot map C onto a single element, hence by the simplicity of C, f_i must be an embedding of C into $M_i \cong \mathcal{L}(P_i)$. Also P_i is infinite dimensional since C contains an infinite chain.

Statement 3: There exists a projective plane Q such that $\mathcal{L}(Q) \in \mathcal{V}$ and Q has at least six points on each line,

By Statement 2, there exists an infinite dimensional nondegenerate projective space P such that $\mathcal{L}(P)$ belongs to $\mathcal{V}$. If every line of P has at least six points, then we can take Q to be any projective plane in P, and $\mathcal{L}(Q) \in \mathcal{V}$ since $\mathcal{L}(Q)$ is a sublattice of $\mathcal{L}(P)$. If the lines of P have less than six points, then by Theorem 3.5, $\mathcal{L}(P)$ is isomorphic to $\mathcal{L}(V, F)$, where V is an infinite dimensional vector space and F is a field with 2, 3 or 4 elements. Let K be a finite field extension of F with $|K| \geq 5$, and let W be a three dimensional vector space over K. As in Section 3.2, $\mathcal{L}(W, K)$ determines a projective plane Q, such that $\mathcal{L}(W, K) \cong \mathcal{L}(Q)$.

Since F is embedded in K, $\mathcal{L}(W, K)$ is a sublattice of $\mathcal{L}(W, F)$, and since V is infinite dimensional, $\mathcal{L}(W, F)$ is isomorphic to a sublattice of $\mathcal{L}(V, F)$. It follows that $\mathcal{L}(Q)$ can be embedded in $\mathcal{L}(P)$, and is therefore a member of $\mathcal{V}$. By construction, every line of Q has at least $|K| + 1 \geq 6$ points.

With the help of these three statements and Lemma 6.19 we can now produce the desired contradiction. Let Q be the projective plane in the last statement. By Statement 2, there exists an infinite dimensional nondegenerate projective space P such that $\mathcal{L}(P) \in$

$\mathcal{V}$ and $\mathcal{L}(Q)$ is isomorphic to a sublattice of $\mathcal{L}(P)$. By the coordinatization theorem (see Section 3.2), there exists a vector space V over a division ring D such that $\mathcal{L}(P)$ is isomorphic to $\mathcal{L}(V, D)$. So $\mathcal{L}(Q)$ is embedded in $\mathcal{L}(V, D) \subseteq \mathcal{S}(V)$ ($=$ the lattice of subgroups of the abelian group V), which implies that the arguesian identity holds in $\mathcal{L}(Q)$. This means Q can be coordinatized in the standard way by choosing any line l in Q and two distinct points a_0, a_∞ on l. The division ring structure is then defined on the set $K = l - \{a_\infty\}$. Here we require only the definition of the addition operation $\oplus$ ([**GLT**] p.208): Choose two distinct points p and q of Q that are collinear with a_0 but are not on l. Given $x, y \in K$, let

$$(1) \qquad u = (x + p)(q + a_\infty) \qquad v = (y + q)(p + a_\infty)$$
$$(2) \qquad x \oplus y = (u + v)l = (u + v)(a_0 + a_\infty)$$

The operation $\oplus$ is independent of the choice of p and q, and $(K, \oplus, a_0)$ is an abelian group.

Any permutation of the points of l induces an automorphism of the quotient $l/0 \subseteq \mathcal{L}(Q)$. Since l has at least six points, we can find $x, y \in K - \{a_0\}$ such that $x \oplus y \neq a_0$. Let α be an automorphism of $l/0$ that keeps a_0, a_∞, x, y fixed and maps $x \oplus y$ to a point $z \neq x \oplus y$. By Lemma 6.19 (ii) there exists an extension L of $\mathcal{L}(Q)$ such that α extends to an automorphism β of L. By Statement 2 and the same argument as above, there exists an abelian group A and an embedding $f : L \hookrightarrow \mathcal{S}(A)$. We claim that $f(x \oplus y)$ is the subgroup of A which satisfies:

$$(3) \quad a \in f(x \oplus y) \quad \text{iff} \quad \begin{array}{l} \text{for some} \quad b \in f(x), \quad c \in f(y) \quad \text{we have} \\ a \ominus b, \ a \ominus c \in f(a_\infty) \quad \text{and} \quad a \ominus b \ominus c \in f(a_0) \end{array}$$

(where $a \ominus b = a \oplus (\ominus b)$ and $\ominus b$ is the additive inverse of b). Indeed, let u, v be as in (1), and assume $a \in f(x \oplus y)$. Since $x \oplus y \leq u + v$, we have $f(x \oplus y) \subseteq f(u + v) = f(u) + f(v)$ ($= \{r \oplus s : r \in f(u), \ s \in f(v)\}$). So there exists $d \in A$ satisfying

$$d \in f(u) \quad \text{and} \quad a \ominus d \in f(v).$$

Since $u \leq x + p$ and $v \leq y + q$, it follows that there exist $b, c \in A$ such that

$$b \in f(x), \quad d \ominus b \in f(p) \quad \text{and} \quad c \in f(y), \quad a \ominus d \ominus c \in f(q).$$

$a \ominus b$ belongs to $f(x \oplus y) + f(x)$ and $f(v) + f(p)$, and since

$$((x \oplus y) + x)(v + p) \leq l(v + p) \leq a_\infty$$

we have that $a \ominus b \in f(a_\infty)$. Similarly $a \ominus c \in f(a_\infty)$ and $a \ominus b \ominus c \in f(a_0)$.

Conversely, assume the right hand side of (3) holds. Since $a_0 \leq p + q$, there exists $d \in A$ such that

$$d \in f(p) \quad \text{and} \quad a \ominus b \ominus c \ominus d \in f(q).$$

Now (1) implies $b + c \in f(u)$ and $a \ominus b \ominus c \in f(v)$, and from (2) we get $a \in f(x \oplus y)$.

Also, in the above argument we can replace f by f' which we define by

$$f'(t) = f(\beta(t)) \quad \text{for all} \quad t \in A.$$

Since f and f' agree on a_0, a_∞, x and y, it follows from (3) that

$$f(x \oplus y) = f'(x \oplus y) = f(z).$$

This is a contradiction, since f is an embedding and $z \neq x \oplus y$. $\square$

6.7 The Day – Ježek Theorem

In this section we will prove the result of Day and Ježek [84]: *if V is a lattice variety that has the amalgamation property and contains the pentagon N, then V must be the variety $\mathcal{L}$ of all lattices.* Together with the preceding result, this implies that T, $\mathcal{D}$ and $\mathcal{L}$ are the only varieties that have the amalgamation property. The proof makes use of the result that $\mathcal{L}$ is generated by the class $\mathcal{B}_F$ of all finite bounded lattices (see Section 2.2). Partial results in this direction had previously been obtained by Berman [81], who showed that if a variety has the amalgamation property and includes N, then it must also include L_3, L_6, L_7, L_9, L_{11} and L_{15}.

The notion of A-decomposability of a finite lattice. This concept was introduced by Slavik [83]. Let L be a finite lattice with L_1 and L_2 proper sublattices of L and $L = L_1 \cup L_2$. L is said to be *A-decomposable by means of L_1 and L_2* (written $L = A(L_1, L_2)$) if whenever (f_1, f_2, L_3) is a completion of L_1 and L_2, then $f = f_1 \cup f_2$ is an embedding of L into L_3. So in a sense $A(L_1, L_2)$ is the smallest completion of L_1 and L_2. In particular, if we let j_i denote the inclusion map of $L_1 \cap L_2$ into L_i ($i = 1, 2$), then $A(L_1, L_2)$ is by definition embeddable into any amalgamation of the diagram $(L_1 \cap L_2, j_1, L_1, j_2, L_2)$. Hence if V is a variety having the amalgamation property, and $L_1, L_2 \in V$, then $A(L_1, L_2) \in V$. This, together with the fact that A-decomposability can be characterized by three easily verifiable conditions on L_1 and L_2, makes it a very useful concept.

For any element $z \in L$ we define $C(z)$ to be the set of all covers of z, and $C^d(z)$ the set of all dual covers of z.

LEMMA 6.21 (Day and Ježek [84]). *Let $L = L_1 \cup L_2$ be a finite lattice with L_1 and L_2 proper sublattices of L. Then L is A-decomposable by means of L_1 and L_2 if and only if L_1 and L_2 also satisfy:*

(1) *$x \in L_i$, $y \in L_j$ and $x \le y$ imply $x \le z \le y$ for some $z \in L_1 \cap L_2$ ($\{i,j\} = \{1,2\}$);*

(2) *$z \in L_1 \cap L_2$ implies $C^d(z) \subseteq L_1$ or $C^d(z) \subseteq L_2$;*

(3) *$z \in L_1 \cap L_2$ implies $C(z) \subseteq L_1$ or $C(z) \subseteq L_2$.*

PROOF. Suppose $L = A(L_1, L_2)$ and let $(f_1, f_2, \mathcal{I}(L_1, L_2))$ be the ideal completion of L_1 and L_2 (see Lemma 6.17). By definition the map

$$f = f_1 \cup f_2 : L \to \mathcal{I}(L_1, L_2) \qquad \text{given by} \qquad f(x) = (x]$$

is a lattice embedding. Let $x \in L_i$, $y \in L_j$ ($\{i,j\} = \{1,2\}$) and $x \le y$. Then $f(x) = (x] \subseteq (y] = f(y)$,hence $x \le_P y$, which implies that there exists a $z \in L_1 \cap L_2$ such that $x \le z \le y$. Therefore (1) holds. Suppose to the contrary that (2) fails. Then there exists $z \in L_1 \cap L_2$ with dual covers $x \in L_1 - L_2$ and $y \in L_2 - L_1$. Clearly $z = x + y$ so $f(z) = (z] = (x] + (y]$. But $(x] \cup (y]$ is already a (L_1, L_2)-ideal, so we should have $(x] \cup (y] = (x] + (y]$. This is a contradiction, since $(z] \ne (x] \cup (y]$. Dually, the existence of the filter completion implies that (3) holds.

Conversely, suppose that (1), (2) and (3) hold, and let (f_1, f_2, L_3) be any completion of L_1 and L_2. We must show that $f = f_1 \cup f_2$ is an embedding of L into L_3. Firstly, $x < y$ implies $f(x) < f(y)$, since if $x, y \in L_i$ this follows from the fact that f_i is an embedding, and if $x \in L_i - L_j, y \in L_j - L_i$ then by (1) there exists a $z \in L_1 \cap L_2$ such that $x \le z \le y$.

Because $x, y \notin L_1 \cap L_2$, we must have $x < z < y$, giving $f(x) < f(z) < f(y)$. This shows that f is one-one and order preserving. To see that f is in fact a homomorphism, requires a bit more work.

Since $f|L_i = f_i$ is a homomorphism, we only have to consider $x \in L_1 - L_2$, $y \in L_2 - L_1$, and show that $f(x + y) = f(x) + f(y)$ ($f(xy) = f(x)f(y)$ follows by duality). We define two maps $\mu_i : L \to L_i$ ($i = 1, 2$) by $\mu_i(u) = \sum \{v \in L_i : v \leq u\}$. Note that the join is taken in L, so $\sum \emptyset = 0_L$. Also, clearly μ_i is orderpreserving, and $\mu_i|L_i$ is the identity map on L_i.

Define two increasing sequences of elements $x_n \in L_1$, $y_n \in L_2$ by $x_0 = x$, $y_0 = y$ and

$$x_{n+1} = x_n + \mu_1(y_n), \qquad y_{n+1} = y_n + \mu_2(x_n).$$

By induction one can easily see that $x_n + y_n = x + y$ and $f(x_n) + f(y_n) = f(x) + f(y)$ for all $n \in \omega$. We show that for some $n = k$ we have $x_k = x + y$ or $y_k = x + y$, then $f(x + y) = f(x_k) + f(y_k) = f(x) + f(y)$ as required.

Suppose $x_n, y_n < x + y$ for all n. Since L is finite, this implies that there exists a k such that $x_{k+1} = x_k$ and $y_{k+1} = y_k$, so by definition $\mu_1(y_k) \leq x_k$ and $\mu_2(x_k) \leq y_k$. We always have $\mu_1(y_k) \leq y_k$ and $\mu_2(x_k) \leq x_k$, hence $\mu_2(x_k), \mu_1(y_k) \leq x_k y_k$. If $x_k y_k \in L_1$ then $x_k y_k \leq \mu_1(y_k)$, so we have $\mu_1(y_k) = x_k y_k \in L_1$. Since $y_k \in L_2$, condition (1) implies that there exists $z \in L_1 \cap L_2$ such that $\mu_1(y_k) \leq z \leq y_k$. But then $z \leq \mu_1(y_k)$, so $z = \mu_1(y_k) = x_k y_k \in L_1 \cap L_2$. Similarly, if $x_k y_k \in L_2$ then we also get $x_k y_k \in L_1 \cap L_2$, hence we actually have

$$\mu_2(x_k) = x_k y_k = \mu_1(y_k) \in L_1 \cap L_2.$$

However $x_k \notin L_1 \cap L_2$ else $\mu_2(x_k) = x_k$ which gives $y_{k+1} = y_k + x_k$, contrary to the initial assumption that $y_n < x + y$ for all n. Similarly $y_k \notin L_1 \cap L_2$, so there exist covers u, v of $x_k y_k$ such that $u \leq x_k$ and $v \leq y_k$. But $\mu_2(x_k) \prec u \leq x_k$ implies $u \in L_1 - L_2$ (else $u \leq \mu_2(x_k)$) and $\mu_1(y_k) \prec v \leq y_k$ implies $v \in L_1 - L_2$. Since this contradicts condition (3), it follows that $x_n = x + y$ or $y_n = x + y$ for some n. $\square$

Two easy consequences of the above characterization are:

COROLLARY 6.22

(i) If $L = A(L_1, L_2)$ and, for $i = 1, 2$, L_i is a sublattice of L'_i, which in turn is a proper sublattice of L then $L = A(L'_1, L'_2)$.

(ii) If $L = [a) \cup (b]$ for some $a, b \in L$ with $0_L \leq a \leq b \leq 1_L$ then $L = A([a), (b])$.

$\mathcal{L}$ is the only nonmodular variety that has the amalgamation property. We also need the following lemma about the lattice $L[I]$ constructed by Day [70] (see Section 2.2).

LEMMA 6.23 Let $I = u/v$ be a quotient in a lattice L, $\theta \in \mathrm{Con}(L)$ and put $J = (u/\theta)/(v/\theta)$. If $I = \bigcup J$, then $L[I]$ is a sublattice of the direct product of L and $L/\theta[J]$.

PROOF. Recall that if ψ, ϕ are two congruences on an algebra A such that $\psi \cap \phi$ is the zero of $\mathrm{Con}(A)$, then A is a subdirect product of A/ψ and A/ϕ. So we need only define ψ and ϕ on $L[I]$ in such a way that $L[I]/\psi$ is a sublattice of L, $L[I]/\phi$ is a sublattice of

$L/\theta[J]$, and $\psi \cap \phi = 0$. Let $\psi = \ker \gamma$, where $\gamma : L[I] \twoheadrightarrow L$ is the natural epimorphism, then $L[I]/\psi$ is of course isomorphic to L. Define ϕ by

$$x\phi y \qquad \text{if and only if} \qquad \gamma(x)\theta\gamma(y) \quad \text{and} \quad \begin{array}{l} x,y \in L - I \quad \text{or} \\ x,y \in I \times \{i\} \quad (i = 1,2). \end{array}$$

With this definition ϕ is a congruence, since h is a homomorphism and $\theta \in \text{Con}(L)$. Moreover, it follows that $I = \bigcup J$,

$$\begin{array}{lll} x \in L - I & \text{implies} & x/\phi = x/\theta \quad \text{and} \\ (x,i) \in I \times \{i\} & \text{implies} & (x,i)/\phi = (x/\theta, i) \quad (i = 0,1) \end{array}$$

whence $L[I]/\phi$ is a subset of $L/\theta[J]$. By examining several cases of meets and joins in $L[I]/\phi$, one sees that it is in fact a sublattice of $L/\theta[J]$.

Suppose now that $x, y \in L[I]$ and $x(\psi \cap \phi)y$. Then $h(x) = h(y)$ and $x, y \in L - I$ or $x, y \in I \times \{i\}$ ($i = 0$ or 1). In all cases it follows that $x = y$, so we have $\psi \cap \phi = 0$ as required. $\qquad\square$

Suppose L is a finite lattice. As in Section 2.2, we let $\kappa(p) = \sum\{x \in L : x \not\geq p$ and $x \geq p_*\}$, where p is any join irreducible of L and p_* is its unique dual cover. Dually we define $\lambda(m) = \prod\{x \in L : x \not\leq m$ and $x \leq m^*\}$ for any meet irreducible $m \in L$.

COROLLARY 6.24 *Let L be a finite semidistributive lattice, and let $I = u/v$ be a quotient in L with $0_L \prec v \leq u \prec 1_L$. Then $L[I]$ is a sublattice of $L \times N$, where N denotes the pentagon.*

PROOF. Clearly v is join irreducible and u is meet irreducible. By semidistributivity, L is the disjoint union of the quotients u/v, $u\kappa(v)/0_L$, $1_L/v + \lambda(u)$ and $\kappa(v)/\lambda(u)$, where the last quotient might be empty if $\kappa(v) \not\geq \lambda(u)$. This defines an equivalence relation θ on L with the quotients as θ-classes. θ is a congruence relation since L is semidistributive, and L/θ is isomorphic to a sublattice of $\mathbf{2} \times \mathbf{2}$. Letting $J = (u/\theta)/(v/\theta) = \{u/\theta\}$ we have $\bigcup J = u/v$. Thus $L/\theta[J]$ is isomorphic to a sublattice of N, and the result follows from the preceding lemma. $\qquad\square$

The following crucial lemma forces larger and larger bounded lattices into any non-modular variety that has the amalgamation property.

LEMMA 6.25 *Let $\mathcal{V}$ be a variety that has the amalgamation property and contains N. If $L \in \mathcal{B}_F \cap \mathcal{V}$ and $v \leq u \in L$, then $L_i = (L \times \mathbf{2})[(u,i)/(v,i)] \in \mathcal{B}_F \cap \mathcal{V}$ $(i = 0,1)$.*

PROOF. It follows from Section 2.2 that all lattices in $\mathcal{B}$ are semidistributive, $\mathbf{2} \in \mathcal{B}_F$, $\mathcal{B}_F$ is closed under the formation of finite products and $L \in \mathcal{B}_F$ implies $L[I] \in \mathcal{B}_F$ for any quotient I of L, so $L \in \mathcal{B}_F$ implies $L_i \in \mathcal{B}_F$ $(i = 0,1)$. We proceed by induction on $|L|$. Assume $i = 1$. If $u = 1_L$ then L_i is a sublattice of $L \times \mathbf{3} \in \mathcal{V}$, hence $L_1 \in \mathcal{V}$. If $u < 1_L$ then there is a co-atom w such that $u \leq w \prec 1$, and $L = (w] \cup [\lambda(w))$. Therefore $L \times \mathbf{2}$ can be pictured as in Figure 6.7 (i).

Let $I = (w,1)/(0,1)$, then $(L \times \mathbf{2})[I]$ is a sublattice of $(L \times \mathbf{2}) \times N$ (by Corollary 6.21), hence a lattice in $\mathcal{V}$. The congruence classes modulo the induced homomorphism $h : (L \times \mathbf{2})[I] \to N$ produce the diagram in Figure 6.7 (ii). Since B_0 is one of these congruence classes, we can double it, again using Day's construction, to obtain a lattice L' as in Figure 6.8.

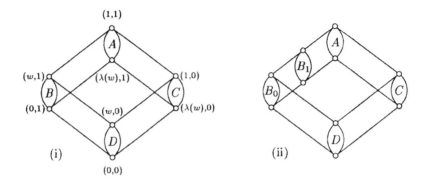

Figure 6.7

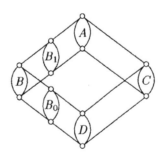

Figure 6.8

Clearly $L' = (L \times \mathbf{2})[I][B_0] \in \mathcal{V}$ by Lemma 6.23. Let J be the quotient u/v considered as lying in the congruence class labeled B in Figure 6.8, and consider the lattice $L'[J] = A \cup B_0 \cup B_1 \cup C \cup D \cup B[J]$. If we define $C_1 = A \cup B_0 \cup B_1 \cup C \cup D$ and $C_2 = B_0 \cup B_1 \cup B[J]$, then we have $L'[J] = A(C_1, C_2)$. Now $C_1 = (L \times \mathbf{2})[I] \in \mathcal{V}$ and $C_2 = A(B_0 \cup B[J], B_1 \cup B[J])$, hence $C_2 \in \mathcal{V}$ if and only if these two lattices belong to $\mathcal{V}$. But $B = w/0$, so $|B| < |L|$, and $B_i \cup B[J] = (B \times \mathbf{2})[(u, i)/(v, i)]$. By induction then $C_2 \in \mathcal{V}$ and this gives $L'[J] \in \mathcal{V}$. Since L_1 is isomorphic to $A \cup B[J] \cup C \cup D$ which is a sublattice of $L'[J]$, L_1 is also in $\mathcal{V}$. The proof for $i = 0$ follows by symmetry. □

THEOREM 6.26 (Day and Ježek [84]). *The only lattice varieties that have the amalgamation property are the variety $\mathcal{T}$ of all trivial lattices, the variety $\mathcal{D}$ of all distributive lattices, and the variety $\mathcal{L}$ of all lattices.*

PROOF. If $\mathcal{V}$ is a nondistributive lattice variety that has the amalgamation property, then $N \in \mathcal{V}$ by Theorem 6.20. The preceding lemma implies that for every $L \in \mathcal{B}_F$ and any $v \le u \in L$, if $L \in \mathcal{V}$ then $L[u/v] \in \mathcal{V}$, since $L[u/v]$ is a sublattice of $(L \times \mathbf{2})[(u, i)/(v, i)]$. It follows Theorem 2.38 that $\mathcal{B}_F \subseteq \mathcal{V}$, and since the finite bounded lattices generate all lattices (Theorem 2.36), we have $\mathcal{V} = \mathcal{L}$. □

6.8 Amal($\mathcal{V}$) for some finitely generated lattice varieties

Let $\mathcal{V}$ be a variety which fails to satisfy the amalgamation property. In this case Amal($\mathcal{V}$) is a proper subclass of $\mathcal{V}$, and it is interesting to find out whether or not Amal($\mathcal{V}$) is an elementary class. In this section we outline the proofs of two results in this direction. They concern the amalgamation classes of finitely generated lattice varieties, and they are surprisingly contrasting with each other: If $\mathcal{V}$ is a finitely generated nondistributive modular lattice variety then Amal($\mathcal{V}$) is not elementary; on the other hand if $\mathcal{V}$ is a variety generated by a pentagon then Amal($\mathcal{V}$) is an elementary class determined by Horn sentences.

Finitely generated modular varieties. We begin with the following:

DEFINITION 6.27 (Albert and Burris [88]).

(i) *Let $\mathcal{V}$ be a variety, and suppose that the diagram (A, f, B, g, C) cannot be amalgamated in $\mathcal{V}$. An obstruction is any subalgebra C' of C such that the diagram (A', f', B, g', C') cannot be amalgamated in $\mathcal{V}$, where $A' = g^{-1}(C')$ and f' and g' are the restrictions of f and g to A'.*

(ii) *Let $\mathcal{V}$ be a locally finite variety. Amal($\mathcal{V}$) is said to have the bounded obstruction property with respect to $\mathcal{V}$ if for every $k \in \omega$ there exists an $n \in \omega$ such that the following holds:*

If $C \in$ Amal($\mathcal{V}$), $|B| < k$ and the diagram (A, f, B, g, C) cannot be amalgamated in $\mathcal{V}$, then there is an obstruction $C' \le C$ such that $|C| < n$.

THEOREM 6.28 (Albert and Burris [88]). *Let $\mathcal{V}$ be a finitely generated variety of finite type. Then Amal($\mathcal{V}$) is elementary if and only if it has the bounded obstruction property.*

Using the preceding theorem, Bergman [89] proved the following result.

THEOREM 6.29 *Let V be a finitely generated nondistributive modular variety. Then $Amal(V)$ is not elementary.*

OUTLINE OF PROOF. Let V be as in the theorem and let L be a finite modular nondistributive lattice which generates V. If S is a subdirectly irreducible member of V then $|S| \leq |L|$, since S is an image of a sublattice of L. Thus S is simple, and since S has a diamond as a sublattice, we have $|S| \geq 5$. Pick S with largest possible cardinality. Let z be the bottom element of S and let $a \in S$ such that a covers z.

Define $B = S \times 2$ and let $f : 2 \hookrightarrow B$ be an embedding with $f(2) = \{(z,0),(a,1)\}$. Then there is $C \in Amal(V)$ and embeddings $g_n : 2 \hookrightarrow C$ such that for each $n \in \omega$, the diagram $(2, f, B, g_n, C)$ cannot be amalgamated in V, and every obstruction has cardinality at least n. (For the details the reader is referred to Bergman [89].) Thus by Theorem 6.28, $Amal(V)$ is not elementary.   □

The variety generated by the pentagon. As before, let $\mathcal{N}$ be the variety generated by the pentagon. The following result appears in Bruyns, Naturman and Rose [a].

THEOREM 6.30 *$Amal(\mathcal{N})$ is an elementary class. It is closed under reduced products and therefore is determined by Horn sentences. Furthermore, if B is an image of $A \in Amal(\mathcal{N})$ and B is a subdirect power of the pentagon then $B \in Amal(\mathcal{N})$.*

The full proof of the above theorem is too long to give here. It requires several definitions and intermediate results. We will list some of them first, and then outline the proof of the theorem. For more details the reader is referred to Bruyns, Naturman and Rose [a].

DEFINITION 6.31

(i) Let θ be a congruence on a lattice A. We shall say that θ is a **2**-congruence if $A/\theta \cong 2$.

(ii) Let A be a subdirect product of lattices $\{A_i : i \in I\}$ and let $B = \bigtimes A_i$. A subdirect representation $A \leq B$ is said to be *regular* if for any kernels θ_i and θ_j of two distinct projections we have $\theta_i|_A \neq \theta_j|_A$.

THEOREM 6.32 Let A be a nontrivial member of $\mathcal{N}$. The following are equivalent:

(i) $A \in Amal(\mathcal{N})$.

(ii) If $A \leq B \in \mathcal{N}$, then every 2-congruence on A can be extended to a 2-congruence on B, and **3** is not an image of A.

(iii) A is a subdirect power of N, and for any regular subdirect representation $f : A \hookrightarrow N^I$ and any homomorphism $g : A \to N$ there is a homomorphism $h : N^I \to N$ such that $g = hf$.

(iv) A is a subdirect power of N, **3** is not an image of A, and if $A \leq N^I$ is a regular subdirect representation, then every 2-congruence on A can be extended to a 2-congruence of N^I.

PROPOSITION 6.33

(i) *Let $B \in \mathcal{N}$, and assume that for any distinct 2-congruences θ_0 and θ_1 on B there is $A \in \mathrm{Amal}(\mathcal{N})$ and embeddings $f_0, f_1 : A \hookrightarrow B$ such that $\theta_0|_{f_0(A)}$ and $\theta_1|_{f_1(A)}$ are two distinct congruences on A. Then $B \in \mathrm{Amal}(\mathcal{N})$.*

(ii) *Let B be an image of $A \in \mathrm{Amal}(\mathcal{N})$ and assume that B is a subdirect power of N. If $B \leq N^I$ is a regular subdirect representation, then every 2-congruence on B can be extended to a 2-congruence on N^I.*

OUTLINE OF THE PROOF OF THEOREM 6.30. We first consider the last statement of the theorem. Let B be an image of $A \in \mathrm{Amal}(\mathcal{N})$. Since **3** is not an image of A we have that **3** is not an image of B. Thus $B \in \mathrm{Amal}(\mathcal{N})$ by Proposition 6.33 (ii) and Theorem 6.32 (iv).

Next we consider direct products. Let $A = \mathsf{X}_{\gamma \in \alpha} A_\gamma$ be a direct product of members of $\mathrm{Amal}(\mathcal{N})$. Without loss of generality we may assume that each A_γ is nontrivial. We use Proposition 6.33 (i) to prove that $A \in \mathrm{Amal}(\mathcal{N})$. Thus we have to show that for any distinct 2-congruences θ_0, θ_1 on A there is a $\gamma \in \alpha$ and embeddings $f_0, f_1 : A_\gamma \hookrightarrow A$ such that $\theta_0|_{f_0(A_\gamma)}$ and $\theta_1|_{f_1(A_\gamma)}$ are two distinct congruences on A. Now if θ_0 and θ_1 are distinct 2-congruences on A, then $A/(\theta_0 \cap \theta_1)$ is isomorphic to either **3** or $\mathbf{2 \times 2}$. In either case we have $u, v, z \in A$ with $u > v > z$ such that

$$(u, v) \in \theta_0, \quad (v, z) \notin \theta_0 \quad \text{and} \quad (u, v) \notin \theta_1, \quad (v, z) \in \theta_1.$$

By Jónsson's Lemma there are congruences ϕ_0, ϕ_1 on A induced by ultrafilters $\mathcal{D}_0$ and $\mathcal{D}_1$ on α such that $\phi_0 \subseteq \theta_0$ and $\phi_1 \subseteq \theta_1$. Defining

$$R = \{\beta \in \alpha : u_\beta > v_\beta\} \qquad S = \{\beta \in \alpha : v_\beta > z_\beta\}$$

we have $R \in \mathcal{D}_0$ and $S \in \mathcal{D}_1$. There are three possible cases:

(i) For some $\gamma \in \alpha$ the set $\{\gamma\}$ belongs to both $\mathcal{D}_0$ and $\mathcal{D}_1$.

(ii) For each $\gamma \in \alpha$ the set $\{\gamma\}$ belongs to neither $\mathcal{D}_0$ nor $\mathcal{D}_1$.

(iii) There exists a $\gamma \in \alpha$ such that $\{\gamma\}$ belongs to one ultrafilter and not the other.

If (i) holds then we can choose u, v, z so that $R = S = \{\gamma\}$, for some $\gamma \in \alpha$. For $\beta \in \alpha$ with $\beta \neq \gamma$ let a_β be an arbitrary but fixed element of A_β. In this case we can have $f_0 = f_1$ so that for $i \in \{0, 1\}$ the embedding $f_i : A_\gamma \hookrightarrow A$ is defined as follows:

For $x \in A_\gamma$ the γth coordinate of $f_i(x) \in A$ is x, and if $\beta \in \alpha$ with $\beta \neq \alpha$ then the βth coordinate of $f_i(x)$ is a_γ.

Suppose now that (ii) holds. Pick any $\gamma \in \alpha$. We have $(R - \{\gamma\}) \in \mathcal{D}_0$ and $(S - \{\gamma\}) \in \mathcal{D}_1$. First observe that since A_γ is nontrivial it is a subdirect power of N (Theorem 6.32 (iii)). Thus there are at least two distinct epimorphisms $r, s : A \to \mathbf{2} = \{0, 1\}$. The embedding $f_0 : A_\gamma \hookrightarrow A$ is defined as follows:

For $x \in A_\gamma$ the γth coordinate of $f_0(x) \in A$ is x. If $\beta \in (R - \{\gamma\})$ then the βth coordinate of $f_0(x)$ is u_β if $r(x) = 1$ and v_β if $r(x) = 0$. For $\beta \in \alpha$ with $\beta \notin (R \cup \{\gamma\})$ the βth coordinate is an arbitrary but fixed element of A_β.

The embedding $f_1 : A_\gamma \hookrightarrow A$ is defined as follows:

For $x \in A_\gamma$ the γth coordinate of $f_1(x) \in A$ is x. If $\beta \in (S - \{\gamma\})$ then the βth coordinate of $f_1(x)$ is v_β if $s(x) = 1$ and z_β if $s(x) = 0$. For $\beta \notin (S \cup \{\gamma\})$ the βth coordinate is an arbitrary but fixed element of A_β.

The case (iii) is a combination of (i) and (ii). For instance if $\{\gamma\} \in \mathcal{D}_0$ and $\{\gamma\} \notin \mathcal{D}_1$, then $(S - \{\gamma\}) \in \mathcal{D}_1$, hence f_0 is defined as in case (i) and f_1 is as in case (ii).

Thus we have shown that every direct product of members of $\text{Amal}(\mathcal{N})$ belongs to $\text{Amal}(\mathcal{N})$. Now if B is a reduced product of members of $\text{Amal}(\mathcal{N})$ then it must be a subdirect power of N (see Bruyns, Naturman and Rose [a] Lemma 0.1.9). On the other hand B is an image of a product A of members of $\text{Amal}(\mathcal{N})$. Since $A \in \text{Amal}(\mathcal{N})$ it follows that $B \in \text{Amal}(\mathcal{N})$. In particular every ultraproduct of members of $\text{Amal}(\mathcal{N})$ belongs to $\text{Amal}(\mathcal{N})$, so that $\text{Amal}(\mathcal{N})$ is elementary (see Yasuhara [74]). It is determined by Horn sentences since it is closed under reduced products (Chang and Keisler [73]).   □

Bibliography

H. Albert and S. Burris
[88] *Bounded obstructions, model companions and amalgamation bases,* Z. Math. Logik. Grundlag. Math. **34** (1988), 109–115.

R. Baer
[52] "Linear Algebra and Projective Geometry," Academic Press, New York (1952).

K. Baker
[69] *Equational classes of modular lattices,* Pacific J. Math. **28** (1969), 9–15.
[74] *Primitive satisfaction and equational problems for lattices and other algebras,* Trans. Amer. Math. Soc. **190** (1974), 125–150.
[77] *Finite equational bases for finite algebras in congruence distributive equational classes,* Advances in Mathematics **24** (1977), 207–243.
[77'] *Some non-finitely based varieties of lattices,* Colloq. Math. **29** (1977), 53–59.

C. Bergman
[85] *Amalgamation classes of some distributive varieties,* Algebra Universalis **20** (1985), 143–166.
[89] *Non-axiomatizability of the amalgamation class of modular lattice varieties,* Order (1989), 49–58.

J. Berman
[81] *Interval lattices and the amalgamation property,* Algebra Universalis **12** (1981), 360–375, MR **82k**: 06007.

G. Birkhoff
[35] *On the structure of abstract algebras,* Proc. Camb. Phil. Soc. **31** (1935), 433–454.
[35'] *Abstract linear dependence and lattices,* Amer. J. Math. **57** (1935), 800–804.
[44] *Subdirect unions in universal algebra,* Bull. Amer. Math. Soc. **50** (1944), 764–768.

P. Bruyns, C. Naturman and H. Rose
[a] *Amalgamation in the pentagon varieties,* Algebra Universalis (to appear).

S. Burris and H. P. Sankappanavar
[81] "A Course in Universal Algebra," Springer-Verlag, New York (1981).

C. Chang and H. J. Keisler
[73] "Model Theory," North-Holland Publ. Co., Amsterdam (1973).

P. Crawley and R. P. Dilworth
[ATL] "Algebraic theory of lattices," Prentice-Hall, Englewood Cliffs, N.J. (1973).

B. A. Davey, W. Poguntke and I. Rival
[75] *A characterization of semi-distributivity,* Algebra Universalis **5** (1975), 72–75.

B. A. DAVEY AND B. SANDS

[77] *An application of Whitman's condition to lattices with no infinite chains*, Algebra Universalis **7** (1977), 171–178.

A. DAY

[70] *A simple solution to the word problem for lattices*, Canad. Math. Bull. **13** (1970), 253–254.

[72] *Injectivity in equational classes of algebras*, Canad. J. Math. **24** (1972), 209–220.

[75] *Splitting algebras and a weak notion of projectivity*, Algebra Universalis **5** (1975), 153–162.

[77] *Splitting lattices generate all lattices*, Algebra Universalis **7** (1977), 163–169.

[79] *Characterizations of lattices that are bounded homomorphic images of sublattices of free lattices*, Canad. J. Math. **31** (1979), 69–78.

A. DAY AND J. JEŽEK

[84] *The amalgamation property for varieties of lattices*, Trans. Amer. Math. Soc., Vol 286 **1** (1984), 251–256.

A. DAY AND B. JÓNSSON

[89] *Non-Arguesian configurations and glueings of modular lattices*, Algebra Universalis **26** (1989), 208–215.

A. DAY AND D. PICKERING

[83] *Coordinatization of Arguesian lattices*, Trans. Amer. Math. Soc., Vol 278 **2** (1983), 507–522.

R. A. DEAN

[56] *Component subsets of the free lattice on n generators*, Proc. Amer. Math. Soc. **7** (1956), 220–226.

R. DEDEKIND

[00] *Über die von drei Moduln erzeugte Dualgruppe*, Math. Ann. **53** (1900), 371–403.

R. P. DILWORTH

[50] *The structure of relatively complemented lattices*, Ann. of Math. (2) **51** (1950), 348–359.

R. FRAÏSSÉ

[54] *Sur l'extension aux relations de quelques properiétés des ordres*, Ann. Sci. École Norm. Sup. (3) **71** (1954), 363–388, MR **16**-1006.

T. FRAYNE, A. C. MOREL AND D. S. SCOTT

[62] *Reduced direct products*, Fund. Math. **51** (1962)/63, 195–228.

R. FREESE

[76] *Planar sublattices of FM (4)*, Algebra Universalis **6** (1976), 69–72.

[77] *The structure of modular lattices of width four, with applications to varieties of lattices*, Memoirs of Amer. Math. Soc. No 181 **9** (1977).

[79] *The variety of modular lattices is not generated by its finite members*, Trans. Amer. Math. Soc. **255** (1979), 277–300, MR **81g**: 06003.

R. FREESE AND J. B. NATION
[78] *Projective lattices*, Pacific J. of Math. **75** (1978), 93–106.
[85] *Covers in free lattices*, Trans. Amer. Math. Soc. (1) **288** (1985), 1–42.

O. FRINK
[46] *Complemented modular lattices and projective spaces of infinite dimension*, Trans. Amer. Math. Soc. **60** (1946), 452–467.

N. FUNAYAMA AND T. NAKAYAMA
[42] *On the distributivity of a lattice of lattice-congruences*, Proc. Imp. Acad. Tokyo **18** (1942), 553–554.

G. GRÄTZER
[66] *Equational classes of lattices*, Duke Math. J. **33** (1966), 613–622.
[GLT] "General lattice theory," Academic Press, New York (1978).
[UA] "Universal Algebra," Second Expanded Edition, Springer Verlag, New York, Berlin, Heidelberg (1979).

G. GRÄTZER AND H. LAKSER
[71] *The structure of pseudocomplemented distributive lattices, II: Congruence extensions and amalgamations*, Trans. Amer. Math. Soc. **156** (1971), 343–358.

G. GRÄTZER, H. LAKSER AND B. JÓNSSON
[73] *The amalgamation property in equational classes of lattices*, Pacific J. Math. **45** (1973), 507–524, MR **51**-3014.

M. D. HAIMAN
[86] *Arguesian lattices which are not linear*, Massachusetts Inst. of Tech. preprint (1986).

M. HALL AND R. P. DILWORTH
[44] *The imbedding problem for modular lattices*, Ann. of Math. **45** (1944), 450–456.

C. HERRMANN
[73] *Weak (projective) radius and finite equational bases for classes of lattices*, Algebra Universalis **3** (1973), 51–58.
[84] *On the arithmetic of projective coordinate systems*, Trans. Amer. Math. Soc. (2) **284** (1984), 759–785.

C. HERRMANN AND A. HUHN
[75] *Zum Wortproblem für freie Untermodulverbände*, Arch. Math. **26** (1975), 449–453.

D. X. HONG
[70] *Covering relations among lattice varieties*, Thesis, Vanderbilt U. (1970).
[72] *Covering relations among lattice varieties*, Pacific J. Math. **40** (1972), 575–603.

J. M. Howie

[62] *Embedding theorems with amalgamation for semigroups*, Proc. London Math. Soc. (3) **12** (1962), 511-534, MR **25**—2139.

A. Huhn

[72] *Schwach distributive Verbände. I*, Acta Sci. Math. (Szeged) **33** (1972), 297–305.

H. Hule

[76] *Über die Eindeutigkeit der Lösungen algebraischer Gleichungssysteme*, Journal für Reine Angew. Math. **282** (1976), 157–161, MR **53**–13080.

[78] *Relations between the amalgamation property and algebraic equations*, J. Austral. Math. Soc. Ser. A **22** (1978), 257–263, MR **58**–5454.

[79] *Solutionally complete varieties*, J. Austral. Math. Soc. Ser. A **28** (1979), 82–86, MR **81i**: 08007.

E. Inaba

[48] *On primary lattices*, J. Fac. Sci. Hokkaido Univ. **11** (1948), 39–107.

P. Jipsen and H. Rose

[89] *Absolute retracts and amalgamation in certain congruence distributive varieties*, Canadian Math. Bull. **32** (1989), 309-313.

B. Jónsson

[53] *On the representation of lattices*, Math. Scand. **1** (1953), 193–206.

[54] *Modular lattices and Desargues theorem*, Math. Scand. **2** (1954), 295–314.

[56] *Universal relational systems*, Math. Scand. **4** (1956), 193–208, MR **20**–3091.

[59] *Arguesian lattices of dimension $n \leq 4$*, Math. Scand. **7** (1959), 133–145.

[60] *Homogeneous universal relational systems*, Math. Scand. **8** (1960), 137–142, MR **23**–A2328.

[60'] *Representations of complemented modular lattices*, Trans. Amer. Math. Soc. **97** (1960), 64–94.

[61] *Sublattices of a free lattice*, Canad. J. Math. **13** (1961), 256–264.

[62] *Algebraic extensions of relational systems*, Math. Scand. **11** (1962), 179–205, MR **27**–4777.

[67] *Algebras whose congruence lattices are distributive*, Math. Scand. **21** (1967), 110–121, MR **38**–5689.

[68] *Equational classes of lattices*, Math. Scand. **22** (1968), 187–196.

[70] *Relatively free lattices*, Coll. Math. **21** (1970), 191–196.

[72] *The class of Arguesian lattices is self-dual*, Algebra Universalis **2** (1972), 396.

[74] *Sums of finitely based lattice varieties*, Advances in Mathematics (4) **14** (1974), 454–468.

[77] *The variety covering the variety of all modular lattices*, Math. Scand. **41** (1977), 5–14.

[79] *On finitely based varieties of algebras*, Colloq. Math. **42** (1979), 255–261.

[80] *Varieties of lattices: Some open problems*, Algebra Universalis **10** (1980), 335–394.

[90] *Amalgamation in small varieties of lattices*, Jour. of Pure and Applied Algebra **68** (1990), 195–208.

B. JÓNSSON AND J. KIEFER
[62] *Finite sublattices of a free lattice*, Canad. J. Math. **14** (1962), 487–497.

B. JÓNSSON AND G. S. MONK
[69] *Representation of primary Arguesian lattices*, Pac. Jour. Math. **30** (1969), 95–139.

B. JÓNSSON AND J. B. NATION
[77] "A report on sublattices of a free lattice," Colloq. Math. Soc. János Bolyai, Contributions to Universal Algebra (Szeged), Vol. 17, North-Holland, Amsterdam (1977), 223–257.

B. JÓNSSON AND I. RIVAL
[79] *Lattice varieties covering the smallest non-modular lattice variety*, Pacific J. Math. **82** (1979), 463–478.

Y. Y. KANG
[87] *Joins of finitely based lattice varieties*, Thesis, Vanderbilt U. (1987).

N. KIMURA
[57] *On semigroups*, Thesis, Tulane University, New Orleans (1957).

E. W. KISS, L. MÁRKI, P. PRÖHLE AND W. THOLEN
[83] *Categorical, algebraic properties. Compendium on amalgamation, congruence extension, endomorphisms, residual smallness and injectivity*, Studia Scientiarum Mathematicarum Hungarica **18** (1983), 19–141.

S. B. KOCHEN
[61] *Ultraproducts in the theory of models*, Ann. Math. Ser. 2 **74** (1961), 221–261.

S. R. KOGALOVSKIĬ
[65] *On a theorem of Birkhoff*, (Russian), Uspehi Mat. Nauk **20** (1965), 206–207.

A. KOSTINSKY
[72] *Projective lattices and bounded homomorphisms*, Pacific J. Math. **40** (1972), 111–119.

R. L. KRUSE
[73] *Identities satisfied by a finite ring*, J. Algebra **26** (1973), 298–318.

J. G. LEE
[85] *Almost distributive lattice varieties*, Algebra Universalis **21** (1985), 280–304.
[85'] *Joins of finitely based lattice varieties*, Korean Math. Soc. **22** (1985), 125–133.

I. V. LVOV
[73] *Varieties of associative rings I, II*, Algebra and Logic **12** (1973), 150–167, 381–393.

R. LYNDON
[51] *Identities in two-valued calculi*, Trans. Amer. Math. Soc. **71** (1951), 457–465.
[54] *Identities in finite algebras*, Proc. Amer. Math. Soc. **5** (1954), 8–9.

F. MAEDA
[51] *Lattice theoretic characterization of abstract geometries*, J. Sci. Hiroshima Univ. Ser.
     A. **15** (1951), 87–96.

A. I. MAL'CEV
[54] *On the general theory of algebraic systems*, (Russian), Mat. Sb. (N.S.) (77) **35**
     (1954), 3–20, MR **17** .

M. MAKKAI
[73] *A proof of Baker's finite basis theorem on equational classes generated by finite ele-
     ments of congruence distributive varieties* , Algebra Universalis **3** (1973), 174–181.

R. MCKENZIE
[70] *Equational bases for lattice theories*, Math. Scand. **27** (1970), 24–38.
[72] *Equational bases and non-modular lattice varieties*, Trans. Amer. Math. Soc. **174**
     (1972), 1–43.

V. L. MURSKIĬ
[65] *The existence in the three-valued logic of a closed class with a finite basis having no
     finite complete system of identities*, (Russian), Dokl. Akad. Nauk. SSSR **163** (1965),
     815–818, MR **32**–3998.

J. B. NATION
[82] *Finite sublattices of a free lattice*, Trans. Amer. Math. Soc. **269** (1982), 311–337.
[85] *Some varieties of semidistributive lattices*, "Universal algebra and lattice theory,"
     (Charlston, S. C., 1984) Lecture Notes in Math. 1149, Springer Verlag, Berlin - New
     York (1985), 198–223.
[86] *Lattice varieties covering $V(L_1)$*, Algebra Universalis **23** (1986), 132–166.

O. T. NELSON
[68] *Subdirect decompositions of lattices of width two*, Pacific J. Math. **24** (1968), 519–523.

B. H. NEUMANN
[54] *An essay on free products of groups with amalgamation*, Philos. Trans. Roy. Soc.
     London Ser. A **246** (1954), 503–554, MR **16**—10.

H. NEUMANN
[67] "Varieties of groups," Springer Verlag, Berlin, Heidelberg (1967).

S. OATES AND M. B. POWELL
[64] *Identical relational in finite groups*, Quarterly J. of Math. **15** (1964), 131–148.

D. PICKERING
[84] *On minimal non-Arguesian lattice varieties*, Ph.D. Thesis, U. of Hawaii (1984).
[a] *Minimal non-Arguesian Lattices*, preprint.

R. S. PIERCE
[68] "Introduction to the theory of abstract algebras," Holt, Rinehart and Winston, New
     York–Montreal, Que.–London (1968), MR **37**–2655.

P. PUDLÁK AND J. TUMA

[80] *Every finite lattice can be embedded in the lattice of all equivalences over a finite set*, Algebra Universalis **10** (1980), 74–95.

I. RIVAL

[76] *Varieties of nonmodular lattices*, Notices Amer. Math. Soc. **23** (1976), A-420.

H. ROSE

[84] *Nonmodular lattice varieties*, Memoirs of Amer. Math. Soc. **292** (1984).

W. RUCKELSHAUSEN

[78] *Obere Nachbarn von $\mathcal{M}_3 + \mathcal{N}$*, Contributions of general algebra, proceedings of the Klagenfurt Conference (1978), 291–329.

O. SCHREIER

[27] *Die Untergruppen der freien Gruppen*, Abh. Math. Sem. Univ. Hamburg 5 (1927), 161–183.

M. SCHÜTZENBERGER

[45] *Sur certains axiomes de la théorie des structures*, C. R. Acad. Sci. Paris **221** (1945), 218–220.

A. TARSKI

[46] *A remark on functionally free algebras*, Ann. of Math. (2) **47** (1946), 163–165.

W. TAYLOR

[72] *Residually small varieties*, Algebra Universalis **2** (1972), 33–53.
[73] *Products of absolute retracts*, Algebra Universalis **3** (1973), 400–401.
[78] *Baker's finite basis theorem*, Algebra Universalis **8** (1978), 191–196.

O. VEBLEN AND W. H. YOUNG

[10] *Projective Geometry*, 2 volumes, Ginn and Co., Bosten (1910).

V. V. VIŠIN

[63] *Identity transformations in a four-valued logic*, (Russian), Dokl. Akad. Nauk. SSSR **150** (1963), 719–721, MR **33**-1266.

J. VON NEUMANN

[60] "Continuous Geometry," Princeton Univ. Press, Princeton, N. J. (1960).

P. WHITMAN

[41] *Free lattices*, Ann. of Math. (2) **42** (1941), 325–330.
[42] *Free lattices. II*, Ann. of Math. (2) **43** (1942), 104–115.
[43] *Splittings of a lattice*, Amer. J. Math. **65** (1943), 179–196.
[46] *Lattices, equivalence relations, and subgroups*, Bull. Amer. Math. Soc. **52** (1946), 507–522.

R. WILLE

[69] *Primitive Länge und primitive Weite bei modularen Verbänden*, Math. Z. **108** (1969), 129–136.

[72] *Primitive subsets of lattices*, Algebra Universalis **2** (1972), 95–98.

M. YASUHARA

[74] *The amalgamation property, the universal-homogeneous models and the generic models*, Math. Scand. **34** (1974), 5–36, MR **51**-7860.

Index